Information and Instructions

This shop manual contains several sections each covering a specific group of wheel type tractors. The Tab Index on the preceding page can be used to locate the section pertaining to each group of tractors. Each section contains the necessary specifications and the brief but terse procedural data needed by a mechanic when repairing a tractor on which he has had no previous actual experience.

Within each section, the material is arranged in a systematic order beginning with an index which is followed immediately by a Table of Condensed Service Specifications. These specifications include dimensions, fits, clearances and timing instructions. Next in order of arrangement is the procedures paragraphs.

In the procedures paragraphs, the order of presentation starts with the front axle system and steering and proceeding toward the rear axle. The last paragraphs are devoted to the power take-off and power lift systems. Interspersed where needed are additional tabular specifications pertaining to wear limits, torquing, etc.

HOW TO USE THE INDEX

Suppose you want to know the procedure for R&R (remove and reinstall) of the engine camshaft. Your first step is to look in the index under the main heading of ENGINE until you find the entry "Camshaft." Now read to the right where under the column covering the tractor you are repairing, you will find a number which indicates the beginning paragraph pertaining to the camshaft. To locate this wanted paragraph in the manual, turn the pages until the running index appearing on the top outside corner of each page contains the number you are seeking. In this paragraph you will find the information concerning the removal of the camshaft.

More information available at haynes.com
Phone: 805-498-6703

J H Haynes & Co. Ltd.
Haynes North America, Inc

ISBN-10: 0-87288-069-9
ISBN-13: 978-0-87288-069-6

Disclaimer

There are risks associated with automotive repairs. The ability to make repairs depends on the individual's skill, experience and proper tools. Individuals should act with due care and acknowledge and assume the risk of performing automotive repairs.

The purpose of this manual is to provide comprehensive, useful and accessible automotive repair information, to help you get the best value from your vehicle. However, this manual is not a substitute for a professional certified technician or mechanic.

This repair manual is produced by a third party and is not associated with an individual vehicle manufacturer. If there is any doubt or discrepancy between this manual and the owner's manual or the factory service manual, please refer to the factory service manual or seek assistance from a professional certified technician or mechanic.

Even though we have prepared this manual with extreme care and every attempt is made to ensure that the information in this manual is correct, neither the publisher nor the author can accept responsibility for loss, damage or injury caused by any errors in, or omissions from, the information given.

SHOP MANUAL

JOHN DEERE

MODELS 50 - 60 - 70
(Gasoline, All-Fuel & LP-Gas)

IDENTIFICATION
Tractor serial number stamped on plate on right side of main case.

INDEX (By Starting Paragraph)

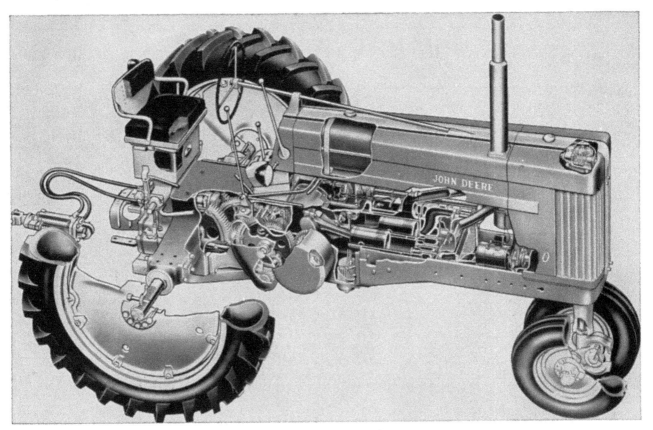

John Deere Model 50

John Deere Model 60

CONDENSED SERVICE DATA

GENERAL	Model 50 (Gasoline)	Model 50 (All-Fuel)	Model 50 (LPG)	Model 60 (Gasoline)	Model 60 (All-Fuel)	Model 60 (LPG)	Model 70 (Gasoline)	Model 70 (All-Fuel)	Model 70 (LPG)
Engine Make	Own	Own	Own	Own	Own	Own	Own	Own	Own
Number Cylinders	2	2	2	2	2	2	2	2	2
Bore—Inches	4 11/16	4 11/16	4 11/16	5½	5½	5½	5⅞	6⅛	5⅞
Stroke—Inches	5½	5½	5½	6¾	6¾	6¾	7	7	7
Displacement—Cubic Inches	190	190	190	321	321	321	379.5	412.5	379.5
Compression Ratio	6.1:1	5.35:1	8.0:1	6.08:1	4.61:1	7.3:1	6.15:1	4.6:1	7.3:1
Compression Pressure at Cranking Speed..	118	91.5	172	123	88	152	115	73	151
Pistons Removed from	Front	Front	Front	Front	Front	Front	Front	Front	Front
Main and Rod Bearings Adjustable?	No	No	No	No	No	No	No	No	No
Cylinders Sleeved?	No	No	No	No	No	No	No	No	No
Forward Speeds	6	6	6	6	6	6	6	6	6
Number of Main Bearings	2	2	2	2	2	2	2	2	2
Generator and Starter Make	D-R	D-R	D-R	D-R	D-R	D-R	D-R	D-R	D-R

TUNE-UP	Model 50 (Gasoline)	Model 50 (All-Fuel)	Model 50 (LPG)	Model 60 (Gasoline)	Model 60 (All-Fuel)	Model 60 (LPG)	Model 70 (Gasoline)	Model 70 (All-Fuel)	Model 70 (LPG)
Valve Tappet Gap (Hot)	0.020	0.020	0.020	0.020	0.020	0.020	0.020	0.020	0.020
Inlet Valve Face Angle	29¾	29¾	29¾	29¾	29¾	29¾	29¾	30	29¾
Inlet Valve Seat Angle	30	30	30	30	30	30	30	30	30
Exhaust Valve Face Angle	44½	44½	44½	44½	44½	44½	44½	45	44½
Exhaust Valve Seat Angle	45	45	45	45	45	45	45	45	45
Ignition Distributor Make	*	*	D-R	D-R	D-R	D-R	D-R	D-R	D-R
Ignition Distributor Model	*	*	1111418	1111558	1111558	1111418	1111558	1111558	1111418
Breaker Gap	**	**	0.022	0.022	0.022	0.022	0.022	0.022	0.022
Distributor Timing—Retard	5°ATDC	TDC	5°ATDC	5°ATDC	TDC	10°ATDC	10°ATDC	TDC	10°ATDC
Distributor Timing—Advanced	20°BTC	25°BTC	10°BTC	20°BTC	25°BTC	5°BTC	15°BTC	25°BTC	5°BTC
Flywheel Mark Indicating:									
Retard Timing	5	TDC	5	5	TDC	10	10	TDC	10
Advanced Timing	20	25	10	20	25	5	15	25	5
Spark Plug Size	18 mm	18 mm	18 mm	18 mm	18 mm	18 mm	18 mm	18 mm	18 mm
Electrode Gap	0.030	0.030	0.030	0.030	0.030	0.030	0.030	0.030	0.030
Carburetor Make	M-S	M-S	Own	M-S	M-S	***	M-S	M-S	***
Model (Early Production)	DLTX75	DLTX73		DLTX81	DLTX72		DLTX82	DLTX85	
Model (Latest Production)	DLTX75	DLTX83		DLTX81	DLTX84		DLTX82	DLTX85	
Float Setting	¾-inch, measured from bowl gasket seat in casting to bottom of float								
Engine Rated RPM	1250	1250	1250	975	975	975	975	975	975
Engine High Idle RPM	1375	1375	1375	1115	1115	1115	1115	1115	1115

SIZES—CAPACITIES—CLEARANCES

(Clearance in Thousandths)

	Model 50 (Gasoline)	Model 50 (All-Fuel)	Model 50 (LPG)	Model 60 (Gasoline)	Model 60 (All-Fuel)	Model 60 (LPG)	Model 70 (Gasoline)	Model 70 (All-Fuel)	Model 70 (LPG)
Crankshaft Journal Diameter (Right)	2.624	2.624	2.624	2.9995	2.9995	2.9995	3.2495	3.2495	3.2495
Crankshaft Journal Diameter (Left)	2.249	2.249	2.249	2.7495	2.7495	2.7495	3.2495	3.2495	3.2495
Pulley Journal Diameter	1.995	1.995	1.995	2.246	2.246	2.246	2.433	2.433	2.433
Crankpin Diameter	2.7492	2.7492	2.7492	2.99925	2.99925	2.99925	3.3743	3.3743	3.3743
Piston Pin Diameter	1.41675	1.41675	1.41675	1.74975	1.74975	1.74975	1.7498	1.7498	1.7498
Valve Stem Diameter	0.434	0.434	0.434	0.4965	0.4965	0.4965	0.4965	0.5585	0.4965
Main Bearings Running Clearance	4-6	4-6	4-6	4-6	4-6	4-6	4.5-6.5	4.5-6.5	4.5-6.5
Rod Bearings Running Clearance	1-4	1-4	1-4	1-4	1-4	1-4	1-4	1-4	1-4
Piston Skirt Clearance	5.5-8.7	5.5-8.7	5.5-8.7	5.1-8.0	5.1-8.0	5.1-8.0	7.0-9.4	6.0-8.9	7.0-9.4
Crankshaft End Play	5-10	5-10	5-10	5-10	5-10	5-10	5-10	5-10	5-10
Cooling System—Gallons	4¾	4¾	4¾	7½	7½	7½	7¾	7¾	7¾
Crankcase Oil—Quarts	7	7	7	8	8	8	11	11	11
Transmission and Differential—Quarts	16	16	16	24	24	24	30	30	30
Powershaft Clutch—Quarts	2.25	2.25	2.25	2.25	2.25	2.25	2.25	2.25	2.25

TIGHTENING TORQUES—FT.-LBS.

	Model 50 (Gasoline)	Model 50 (All-Fuel)	Model 50 (LPG)	Model 60 (Gasoline)	Model 60 (All-Fuel)	Model 60 (LPG)	Model 70 (Gasoline)	Model 70 (All-Fuel)	Model 70 (LPG)
Cylinder Head	104	104	104	150	150	150	208	208	208
Connecting Rod Nuts	85	85	85	105	105	105	85	85	85
Main Bearing Cap Screws	100	100	100	150	150	150	150	150	150
Spark Plugs	35	35	35	35	35	35	35	35	35
Cylinder Block to Main Case	167	167	167	167	167	167	208	208	208
Flywheel Bolts	150	150	150	275	275	275	275	275	275
Brake Pedal Shaft Nut	63	63	63	63	63	63	63	63	63
Rear Axle Housing to Case Screws	150	150	150	150	150	150	150	150	150
Rockshaft Cylinder Cap Screws	150	150	150	150	150	150	150	150	150
Remote Cylinder End Cap	85	85	85	85	85	85	85	85	85
Powershaft Clutch Housing	50	50	50	50	50	50	50	50	50

*Models prior to Ser. No. 5016500 were factory equipped with a Wico XB4023 battery ignition unit which can be replaced with the late production Delco-Remy No. 1111558.

**Wico, 0.015. Delco-Remy, 0.022.

***Single induction, Century; dual induction, own.

FRONT SYSTEM—TRICYCLE TYPE

Models 50 and 60 dual wheel tricycle type tractors are available with the standard one-piece pedestal shown in Fig. JD1000, and the wheels are either the conventional knuckle mounted type or the "Roll-O-Matic" type as shown in Fig. JD1001. Model 70, as well as models 50 and 60 tricycle type tractors, are available with a convertible type, two-piece pedestal as shown in Fig. JD1002. The available convertible front systems include a fork mounted single front wheel, dual *wheels of the conventional knuckle mounted type or dual wheels of the "Roll-O-Matic" type. Refer to Fig. JD1003.*

STEERING SPINDLE AND PEDESTAL
Models 50-60-70
(Manual Steering)

1. **R&R AND OVERHAUL.** To remove the steering spindle (16—Figs. JD1001 and 1002) and knuckle assembly, first remove grille, steering wheel and the steering wheel Woodruff key. Unbolt baffle plate from the radiator top tank. Remove cap screws retaining the steering worm housing to the pedestal and bump the housings apart as shown in Fig. JD1004. Withdraw the steering worm and housing assembly from tractor. Remove cover (or cup plug on late models) from top of pedestal and remove cap screw or nut retaining steering gear to spindle. Raise front of tractor and using a knocker tool or brass drift, bump spindle down and out steering gear. Withdraw gear and save the adjusting washers which are located under the gear. Continue to raise front of tractor and withdraw vertical spindle and knuckle assembly from below. Note: On models equipped with the convertible, two-piece pedestal, it is advisable, due to the additional weight, to unbolt the lower pedestal extension and wheels assembly from the steering spindle, prior to removal.

The spindle lower bearing (5—Figs.

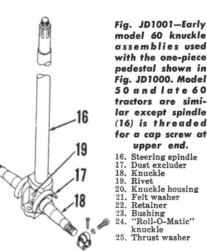

Fig. JD1001—Early model 60 knuckle assemblies used with the one-piece pedestal shown in Fig. JD1000. Model 50 and late 60 tractors are similar except spindle (16) is threaded for a cap screw at upper end.

16. Steering spindle
17. Dust excluder
18. Knuckle
19. Rivet
20. Knuckle housing
21. Felt washer
22. Retainer
23. Bushing
24. "Roll-O-Matic" knuckle
25. Thrust washer

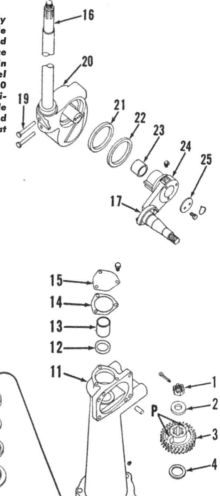

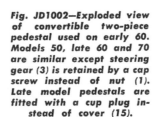

Fig. JD1000—Exploded view of the standard one-piece pedestal on early model 60. Model 50 and late 60 tractors are similar except steering gear is retained to spindle by a cap screw instead of nut (1). The one-piece pedestal is used in conjunction with the knuckle assemblies shown in Fig. JD1001. Late model pedestals are fitted with a cup plug instead of cover (15).

1. Nut or cap screw	10. Spacer (60 only)
2. Washer	11. Pedestal (one-piece)
3. Steering gear	12. "O" ring
4. Adjusting washers	13. Bushing
5. Lower bearing	14. Gasket (early models)
7. Cork washer	15. Cover (early models), cup plug (late models)
8. Gasket	
9. Cap	

Fig. JD1002—Exploded view of convertible two-piece pedestal used on early 60. Models 50, late 60 and 70 are similar except steering gear (3) is retained by a cap screw instead of nut (1). Late model pedestals are fitted with a cup plug instead of cover (15).

1. Nut or cap screw
2. Washer
3. Steering gear
4. Adjusting washers
5. Lower bearing
6. Washer
7. Cork washer
11. Pedestal (two-piece)
12. "O" ring
13. Bushing
14. Gasket (early models)
15. Cover (early models), cup plug (late models)
26. Retainer

JD1000 or 1002) can be removed by using a suitable puller after removing retainer (26) or cap (9).

2. Inside diameter of a new spindle upper bushing (13—Figs. JD1000 or 1002) is 1.500-1.503 and the diameter of the steering spindle upper bearing surface should be 1.497-1.498. When installing the bushing, the beveled edge should be toward bottom of pedestal, the split should be toward the steering worm and top of bushing should be flush with top of bushing bore. Check the installed bushing to make certain that the steering spindle has the recommended clearance of 0.002-0.006 in the bushing. If there is evidence of oil leakage to the lower part of the pedestal, renew "O" ring (12) which is located just below the bushing (13).

3. When installing the steering spindle, observe the following:

(a) Be careful not to damage the "O" ring (12) when lowering front of tractor onto spindle.

(b) Vary the number of spacer washers (4) to give the spindle an end play of 0.004-0.040.

(c) With front wheel (or wheels) pointing straight ahead, the hub projections (P) on steering gear should be parallel to center line of tractor.

"ROLL-O-MATIC"
Models 50-60-70

4. **OVERHAUL.** Regardless of whether the "Roll-O-Matic" unit is used with the standard one-piece pedestal on models 50 and 60 or the two-piece, convertible type pedestal on models 50, 60 and 70, the overhaul procedure is identical and can be accomplished without removing the unit from the tractor.

Support front of tractor and remove wheel and hub units. Remove knuckle caps (45—Fig. JD1003C). Unbolt and remove thrust washers (25). Pull knuckle and gear units from housing and remove felt washer (21). Check the removed parts against the values which follow:

Knuckle Bushing Inside Diameter
    Model 50 . . . . 1.623-1.625
    Models 60-70 . . . . 1.873-1.875

Fig. JD1004—Models 50, 60 and 70 steering (worm) housing separated from pedestal. Shims (76) are used to adjust the backlash between the worm and steering gear.

11. Pedestal	77. Worm housing
73. Worm shaft	80. Bearing housing

Knuckle Shaft Outside Diameter
    Model 50 . . . . 1.620-1.622
    Models 60-70 . . . . 1.870-1.872
Thickness of Thrust Washers (25)
    Models 50-60-70 . . . .   0.156

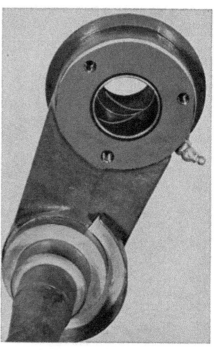

Fig. JD1005—Models 50, 60 and 70 "Roll-O-Matic" knuckle, showing the proper installation of bushings. Notice that open end of oil grooves in bushings is toward the 1/32-1/16 inch gap between the bushings.

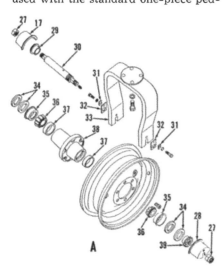

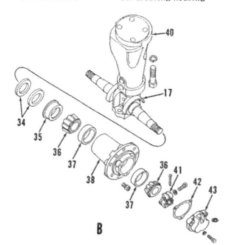

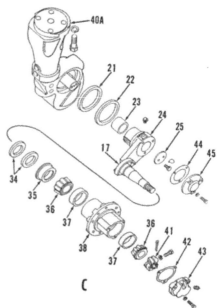

Fig. JD1003—Models 50, 60 and 70 tricycle type front systems used with the convertible, two-piece pedestal shown in Fig. JD1002.
A. Single front wheel. B. Conventional knuckle mounted dual front wheels. C. "Roll-O-Matic."

17. Dust excluder	25. Thrust washer	33. Yoke	40. Pedestal extension	41. Nut
21. Felt washer	27. Nut	34. Felt washers	(except "Roll-O-	42. Gasket
22. Retainer	29. Bearing spacer	35. Retainer	Matic")	43. Cap
23. Bushing	30. Axle	36. Bearing cone	40A. Pedestal exten-	44. Gasket
24. "Roll-O-Matic"	31. Nut lock plate	37. Bearing cup	sion ("Roll-O-	45. Cap
knuckle	32. Axle lock plate	38. Hub	Matic")	

Fig. JD1006—When assembling the 50, 60 and 70 "Roll-O-Matic" unit make certain that gears are meshed so that timing marks (M) are in register.

Install knuckle bushings (23) with open end of oil groove toward gap between bushings as shown in Fig. JD1005. When the bushings are properly installed, there should be a gap of 1/32-1/16 inch between the bushings to allow grease from the fittings to enter grooves in bushings.

Soak felt washers (21—Fig. JD1003C) in engine oil prior to installation. Install one of the knuckles so that wheel spindle extends behind the vertical steering spindle. Pack the "Roll-O-Matic" unit with wheel bearing grease and install the other knuckle so that timing marks on gears are in register as shown in Fig. JD1006.

FRONT SYSTEM—AXLE TYPE

STEERING KNUCKLES
Models 50-60-70

5. **R&R AND OVERHAUL.** Before removing knuckle, raise front of tractor and check end play of knuckle post in the axle knee. If end play exceeds 0.036, renew thrust washers (56—Fig. JD1007) when reassembling.

To remove either knuckle, remove wheel and hub assembly, disconnect drag link from steering arm and remove nut retaining steering arm to top of knuckle post. Using a knocker tool or brass drift, drive knuckle post free of steering arm. Inside diameter of new knuckle post bushings should be 1.504-1.506 and the knuckle post diameter should be 1.494-1.495.

When installing new bushings, press them in until flush with top and bottom of knee and make certain that oil groove in each bushing is in register with fittings in knee. Ream the bushings after installation, if necessary to give the knuckle post the recommended clearance of 0.009-0.012 in the bushings.

When reassembling, install dust shield (55) on knee. Install thrust washers (56) on knuckle post and install knuckle and post assembly into knee, making certain that dowel pin engages holes in thrust washers. Install steering arm (49) so that centerline of steering arm is in a plane 90 degrees to centerline of wheel knuckle.

DRAG LINKS AND TOE-IN
Models 50-60-70

6. An adjustable ball socket is fitted to both ends of each drag link. Drag link ends should be adjusted so they have no end play, yet do not bind.

With front wheels pointing straight ahead, toe-in should be 1/8-3/16 inch. If the adjustment is not as specified, disconnect drag link ends from steering arms and adjust the length of each drag link, an equal amount, until proper adjustment is obtained.

AXLE PIVOT PINS AND BUSHINGS
Models 50-60-70

7. To renew the axle pivot pins and bushings, support front of tractor and disconnect the center steering arm

Fig. JD1007—Models 50, 60 and 70 front axle exploded view. The center member and pivot bracket are shown in Fig. JD1008.

17. Dust excluder
46. Extension (some models)
47. Knee
48. Washer
49. Steering arm
50. Washer
51. Bushing
52. Drag link cover
53. Dowel pin
54. Knuckle & post
55. Dust shield
56. Thrust washers
57. Screw plug
58. Ball stud bearing
59. Drag link end
60. Drag link rod
61. Drag link sleeve
62. Center steering arm

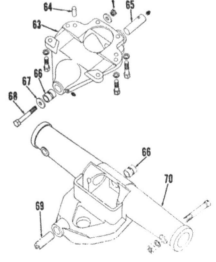

Fig. JD1008—Models 50, 60 and 70 axle center member and pivot bracket.

63. Pivot bracket	67. Washer
64. Dowel	68. Bolt
65. Front pivot pin	69. Rear pivot pin
66. Bushing	70. Axle center member

from the steering spindle. Unbolt axle pivot bracket from front end support and roll axle, pivot bracket and wheels assembly forward and away from tractor. Remove bolt (68—Fig. JD1008) and withdraw pivot bracket from axle. Inside diameter of new pivot pin bushings is 1.504-1.506 and diameter of new pivot pins is 1.494-1.495.

When installing a new bushing in axle or pivot bracket, press or drive bushing in until bushings are flush with castings and oil groove in each bushing is in register with grease fittings in axle and pivot bracket. Ream the bushings after installation, if necessary, to provide a clearance of 0.009-0.012 for the pivot pins. Pivot pins can be driven from axle and pivot bracket and new ones can be driven or pressed in. When properly installed, pivot pins should extend 2¼ inches from axle and pivot bracket.

Reassemble and reinstall the axle and pivot bracket assembly by reversing the removal procedure.

STEERING SPINDLE AND PEDESTAL
Models 50-60-70
(Manual Steering)

8. **R&R AND OVERHAUL.** To remove the steering spindle (16—Fig. JD1002), first remove grille, steering wheel and the steering wheel Woodruff key. Unbolt baffle plate from radiator top tank. Remove cap screws retaining the steering worm housing to the pedestal and bump the housings apart as shown in Fig. JD1004. Withdraw the steering worm and housing assembly from tractor. Support front of tractor and disconnect center steering arm from the steering spindle. Unbolt axle pivot bracket from front end support and roll axle, pivot bracket and wheels assembly forward and away from tractor. Remove cover (or cup plug on late models) from top of pedestal and remove cap screw or nut retaining steering gear to spindle. Using a knocker tool or brass drift, bump spindle down and out of steering gear. Withdraw gear and save the adjusting washers which are located under the gear.

The spindle lower bearing (5—Fig. JD1002) can be removed by using a suitable puller after removing retainer (26).

Inside diameter of a new spindle upper bushing (13) is 1.500-1.503 and the diameter of the steering spindle upper bearing surface should be 1.497-1.498. When installing the bushing,

the beveled edge should be toward bottom of pedestal, the split should be toward the steering worm and top of bushing should be flush with top of bushing bore. Check the installed bushing to make certain that the steering spindle has the recommended clearance of 0.002-0.006 in the bushing. If there is evidence of oil leakage to the lower part of the pedestal, renew "O" ring (12) which is located just below the bushing (13).

When installing the steering spindle, observe the following:

(a) Be careful not to damage the "O" ring (12) when installing the steering spindle.

(b) Vary the number of spacer washers (4) to give the spindle an end play of 0.004-0.040.

(c) With steering spindle positioned so that when the center steering arm is connected and the front wheels are pointing straight ahead, the hub projections (P) on steering gear should be parallel to center line of tractor.

MANUAL STEERING SYSTEM

Models 50-60-70

For the purposes of this discussion, the steering mechanism will include the steering worm and shaft and the steering gear. For R&R and overhaul of the steering spindle, refer to paragraphs 1 or 8.

9. **ADJUSTMENT.** Three adjustments are provided on the steering mechanism: (1) steering spindle shaft end play (2) steering (worm) shaft end play (3) backlash between the worm and steering gear.

10. **STEERING SPINDLE SHAFT END PLAY.** The desired steering spindle shaft end play of 0.004-0.040 is controlled by adjusting washers (4—Figs. JD1000 or 1002) which are located under the steering gear (3). To make the adjustment, it is necessary to remove the steering gear from tractor as outlined in paragraph 14.

11. **STEERING (WORM) SHAFT END PLAY.** The desired steering (worm) shaft end play of 0.001-0.004 is controlled by shims (81 — Fig. JD1009) which are located under the worm shaft front bearing housing. To check, attach dial indicator to steering shaft with indicator contact button resting on the steering shaft rear support. While holding steering wheel to keep it from turning, move front wheel (or wheels) back and forth without turning steering wheel and note the end play on the dial indicator. If end play is not as specified, remove the grille and worm shaft front bearing housing and add or remove shims until proper adjustment is obtained. Thick, medium and thin shims are available.

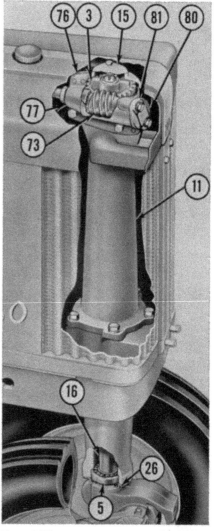

Fig. JD1009—Cut-away view of models 50, 60 and 70 steering mechanism.

3. Steering gear	73. Worm
11. Pedestal	76. Shims
15. Cover (early models) cup plug (late models)	77. Worm housing
	80. Bearing housing
	81. Shims

12. BACKLASH. The desired backlash between the steering gear and worm is ½-1 inch when measured at rim of steering wheel, and is controlled by shims (76—Fig. JD1009) which are located between the pedestal and worm housing. If backlash is not as specified, remove grille and unbolt baffle plate from the radiator top tank. Unbolt worm housing from pedestal, bump housings apart and add or remove shims (76) until backlash is as specified. Shims are available in thicknesses of 0.005 & 0.015. If excessive backlash cannot be eliminated, it will be necessary to renew the worm and steering gear; or, reposition the steering gear on the steering spindle so as to bring unworn teeth into mesh. Refer to paragraph 14.

13. **OVERHAUL.** The steering mechanism can be overhauled without removing pedestal from tractor, as follows:

14. STEERING GEAR. To remove the steering gear (3—Figs. JD1000 or 1002), first remove grille, steering wheel and the steering wheel Woodruff key. Unbolt baffle plate from radiator top tank and remove pedestal cover (or cup plug on late models). Unbolt worm housing from pedestal and bump the housings apart. Withdraw wormshaft and housing assembly from tractor. On axle type trac-

tors, disconnect center steering arm from steering spindle and unbolt axle pivot bracket from front end support. On all models, raise front of tractor, remove nut or cap screw retaining steering gear to spindle and using a knocker tool or brass drift, bump spindle down and out of steering gear.

Reinstall the steering gear by reversing the removal procedure and observe the following:

(a) If the same gear is being reinstalled, position the gear so that unworn teeth will mesh with the worm.

(b) Vary the number of spacer washers (4) to give the spindle an end play of 0.004-0.040.

(c) With front wheel (or wheels) pointing straight ahead, the hub projection (P) on steering gear should be parallel to center line of tractor.

15. STEERING WORM. Overhaul of the steering (worm) shaft and components is accomplished as follows: Remove grille, steering wheel and the steering wheel Woodruff key. Remove the worm shaft front bearing housing (80—Fig. JD1010) and on models so equipped, loosen the packing nut at rear of worm shaft housing. Turn worm shaft forward and out of housing. The worm shaft housing can be removed from pedestal at this time.

The need and procedure for further disassembly is evident after an examination of the unit. Inspect packing (75) on models so equipped or the spring loaded "O" ring seal on some models, and renew if damaged.

When reassembling, adjust the worm shaft end play and the gear unit backlash as outlined in paragraphs 11 and 12.

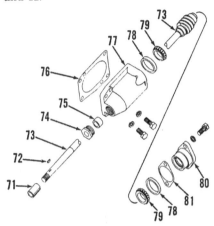

Fig. JD1010—Exploded view of models 50 and 60 steering (worm) shaft and associated parts. Shims (76) control backlash between worm and steering gear. Shims (81) control end play of worm shaft. Late model 60 tractors and all model 70 tractors are equipped with a spring loaded "O" ring seal instead of packing (75) and nut (74). The "O" ring is held in place by a spring on the worm shaft.

71. Bushing
72. Woodruff key
73. Worm shaft
74. Packing nut (some models)
76. Shims
77. Worm housing
78. Bearing cup
79. Bearing cone
80. Bearing housing

POWER STEERING SYSTEM

Note: The maintenance of absolute cleanliness of all parts is of utmost importance in the operation and servicing of the hydraulic power steering system. Of equal importance is the avoidance of nicks or burrs on any of the working parts.

LUBRICATION

16. It is recommended that the power steering system be drained (but not flushed) once a year or every 1,000 hours. To drain the system, turn the steering wheel to the extreme left position and remove the grille. Remove the reservoir drain plug and allow system to drain. To remove the additional oil remaining in the steering cylinder, disconnect the front oil line from the control valve, pivot the line forward and turn the steering wheel to the extreme right position.

Refill the system with five U.S. quarts of John Deere power steering oil No. AF 2235 R.

TROUBLE SHOOTING

17. The following paragraphs outline the possible causes and remedies for troubles in the power steering system.

17A. DRIFTING TO EITHER SIDE could be caused by:

1. Valve housing not in correct relation to worm shaft housing. Center the valve housing as in paragraph 23.

17B. HARD STEERING could be caused by:

1. Insufficient volume of oil flowing to steering valve from flow control valve. Adjust the flow control valve as in paragraph 22.

2. Excessive tension on the centering cam spring. Adjust the spring as in paragraph 29A.

3. Insufficient end play or binding in the worm shaft bearings. See paragraph 27.

4. Insufficient backlash between the steering worm and gear. See paragraph 29.

5. Excessive leakage past the steering vane seals in cylinder. To check, remove grille, disconnect the front oil line from steering valve, pivot the oil line forward and tighten the lower oil line connection. With the engine running at fast idle, hold the steering wheel to the extreme right turn position and measure the leakage from the oil line as shown in Fig. JD1012. Oil leakage should not exceed 1 quart in ½ minute. If leakage is excessive, renew the vane seals. Refer to paragraph 29B.

6. Binding in the rear steering shaft support bushing.

7. Pedestal improperly aligned. See paragraph 28.

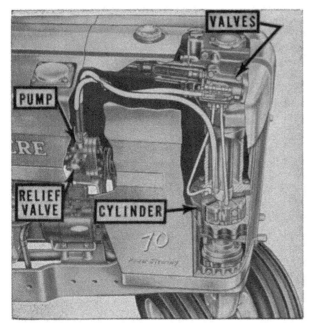

Fig. JD1011—Cut-away of John Deere 70 tractor, showing the installation of the power steering components. On later models, the relief valve is located in the valve housing.

bent steering spindle. Refer to paragraph 29B.

SYSTEM OPERATING PRESSURE

Series 50 Prior Ser. No. 5029200
Series 60 Prior Ser. No. 6054500
Series 70 Prior Ser. No. 7031300

18. On early models, the system relief valve was mounted in the hydraulic pump housing as shown in Fig. JD1011. To check the system operating pressure, remove grille and install a suitable pressure gage (at least 1500 psi capacity) and shut-off valve in series with the pump pressure line as shown in Fig. JD1013. Start engine, close the shut-off valve and observe the highest pressure reading which should be 1170-1210 psi.

CAUTION: Operate system with the shut-off valve closed ONLY long enough to obtain a pressure reading, then open the valve. Pump could be damaged if operated at relief pressure for too long a period.

If the operating pressure is not as specified, vary the number of washers (19—Fig. JD1014) which are located between the relief valve spring and the cover. If the addition of washers will not increase the pressure to within the specified limits, look for a faulty pump.

8. Insufficient pump pressure. Refer to paragraph 18 or 19.
9. Insufficient oil in system.
10 Foaming oil in system. Refer to paragraph 17D.

17C. EXCESSIVE INSTABILITY OF FRONT WHEELS. This condition is often referred to as shimmy or flutter. In some cases, flutter cannot be entirely eliminated, but the unit can be adjusted to the point where it is not objectionable. Possible causes of instability are:

1. Excessive volume of oil flowing from flow control valve to steering valve. Adjust the flow control valve as in paragraph 22.
2. Actuating sleeve set screw too loose in helix slot. Refer to paragraph 27.
3. Worn point on actuating sleeve set screw and/or damaged helix slot in steering worm shaft.
4. Insufficient tension on cam spring. Refer to paragraph 29A.

5. Excessive end play in the steering worm shaft. See paragraph 27.
6. Excessive backlash between the steering worm and gear. Refer to paragraph 29.
7. Unbalanced front wheels.
8. Loose or worn front wheel bearings.
9. Loose or worn "Roll-O-Matic" assembly.
10. Use of front wheel weights rather than front end weights.

17D. OVERFLOW OR FOAMING OF OIL could be caused by:
1. Air leak in system. To check, apply a light coat of oil to sealing surfaces and observe for leaks.
2. Wrong type oil. Use only John Deere power steering oil AF2235 R.

17E. LOCKING could be caused by:
1. Scored worm, or worm bearings adjusted too tight. Refer to paragraph 27.
2. Steering valve arm interference in groove of actuating sleeve.
3. Steering gear loose on vertical spindle. Tighten the cap screw to a torque of 190 Ft-Lbs.
4. Bent steering spindle or scored steering spindle bearings. Refer to paragraph 29B.
5. Loose or broken movable vane retaining cap screws. Screws should be tightened to a torque of 208 Ft.-Lbs. Refer to paragraph 29B.
6. Insufficient clearance between actuating sleeve and the steering worm shaft or cam. Refer to paragraph 29A.

17F. VARIATION IN STEERING EFFORT WHEN TURNING IN ONE DIRECTION could be caused by a

Series 50 After Ser. No. 5029199
Series 60 After Ser. No. 6054499
Series 70 After Ser. No. 7031299

19. On late models, the system relief valve is mounted in the steering valve housing as shown in Fig. JD-1019. To check the system operating pressure, remove grille and install a suitable pressure gage (at least 1500 psi capacity) in series with the pump pressure line in a manner similar to that shown in Fig. JD1013, but do not install the shut-off valve which is shown in the illustration. Cramp the front wheels to the extreme right or extreme left position, start engine and

Fig. JD1012—Checking for leakage past the power steering cylinder vane.

Fig. JD1013—Pressure gage and shut-off valve installation for checking the power steering system operating pressure on models where the relief valve is located within the pump housing.

observe the highest pressure reading which should be 1170-1210 psi.

If the operating pressure is not as specified, vary the number of washers (19—Fig. JD1019). If the addition of washers will not increase the pressure to within the specified limits, look for a faulty pump.

PUMP

20. **REMOVE AND REINSTALL.** To remove the power steering pump, first remove the grille and proceed as follows: Disconnect the steering shaft coupling and remove the steering shaft and hood. Drain the power steering reservoir and remove the pump pressure and suction oil lines. Loosen and disconnect the fan belt. Unbolt the fan shaft tube flange from pump and pump from mounting bracket.

Move the pump and fan assembly forward until the fan shaft coupling is just exposed. While holding the coupling rearward, move the pump assembly forward until the pump shaft is disengaged from the coupling.

Note: If coupling is not held rearward while pump is moved forward, the fan shaft may come out of the rear coupling. If the shaft comes out of the rear coupling, it is diffficult to replace.

Move rear of pump toward right side of tractor, tip top of pump toward right side and withdraw the pump and fan assembly from left side of tractor.

Note: In some individual cases it is more convenient and often time will be saved by removing the fan blades before unbolting the pump from its mounting bracket.

Install the pump by reversing the removal procedure and fill the pump with oil before connecting the oil lines.

21. **OVERHAUL.** With the pump removed from tractor, thoroughly clean the exterior surfaces to remove any accumulation of dirt or other foreign material. Remove the pump housing retaining cap screws and using a plastic hammer, bump the pump housing from its locating dowels. Remove the pump body, follower gear and drive gear. Refer to Fig. JD1015. Extract the drive gear Woodruff key (2) from

the fan shaft. On early model pumps, remove the relief valve cover, spring and relief valve, being careful not to lose the pressure adjusting washers located between the relief valve cover and the spring.

Place the pump cover and fan assembly on a press, depress the fan against spring pressure as shown in Fig. JD1016 and remove the split cone locks (9) and keeper (8). Remove assembly from press and withdraw the fan, pulley, spring and drive parts from the shaft. Press the pump drive shaft rearward out of pump cover, extract the bearing retaining snap ring and remove bearing from the cover. Inspect all parts and renew any which are questionable. Bushings in pump body and cover should be pressed in with a piloted arbor. Oil seal in pump cover should be pressed in not more than $\frac{1}{16}$-inch below bottom of bearing bore.

To reassemble the pump, install the cover bearing and snap ring. Press the pump shaft into bearing from gear side of cover until shoulder on shaft seats against the bearing race. Install the two assembly cap screws in cover, then install the driving pulley, spring and fan parts in the sequence shown in Fig. JD1014. Install Woodruff key in the drive shaft and slide the drive gear into position. Drive gear must float on shaft and not bind on the Woodruff key. Install the follower gear, coat the mating surfaces of the pump cover and body with shellac and install the body so that edges of body are concentric with edges of cover. Coat the mating surfaces of the pump body and housing with shellac, install the housing and tighten the assembly cap screws securely. On early models so equipped, install the relief valve, spring, adjusting washers and the relief valve cover.

STEERING VALVES

22. **ADJUST FLOW CONTROL VALVE.** Turning the flow control valve adjusting screw inward will cause faster, easier steering action; but, the increase in steering speed can

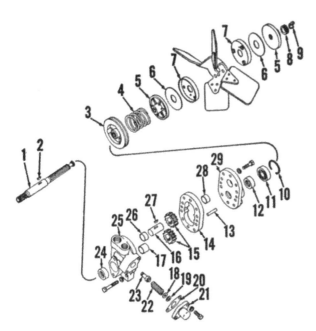

Fig. JD1014 — Exploded view of power steering pump and pressure relief valve used on early models. The pump used on later models is similar, except the relief valve is mounted in valve housing.

1. Fan shaft
5. Friction disc
6. Friction washer
7. Drive cup
8. Fan keeper
9. Locks
10. Snap ring
11. Bearing
12. Oil seal
13. Dowel pin
14. Pump body
15. Pumping gears
16. Idler gear shaft
17. Bushing
18. Washer
19. Shim washers
20. Gasket
21. Relief valve cover
22. Relief valve spring
23. Relief valve
24. Oil seal
25. Pump housing
26. Bushing
27. Woodruff key
28. Bushing
29. Pump cover

Fig. JD1015 — Partially disassembled view of power steering pump. Be sure to remove Woodruff key (2) before attempting to remove the fan shaft (1).

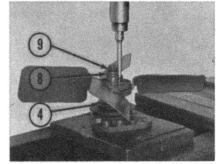

Fig. JD1016 — Compressing the fan shaft spring (4) to permit the removal of locks (9) and keeper (8).

result in a decrease of front wheel stability (increased tendency of front wheels to flutter).

To make the adjustment, operate the steering system until the oil is at normal operating temperature; then, turn the adjusting screw in until the fastest turning speed is obtained without causing an objectionable amount of front wheel flutter. Refer to Fig. JD1017.

Note: Front wheel instability (flutter) can also be caused by improper adjustment of the steering worm and sleeve assembly. Refer to paragraph 27.

23. CENTERING VALVE HOUSING. Housing (40—Fig. JD1019) is properly centered when the effort required to turn the steering wheel in one direction is exactly the same as the effort required to turn the steering wheel in the other direction. The adjustment is made by loosening the valve housing retaining cap screws, tapping the housing either way as required and then tightening the cap screws.

If the wheel turns harder to the left, tap the housing rearward. Conversely, if the wheel turns harder to the right, tap the valve housing forward.

24. REMOVE AND REINSTALL. To remove the valve housing, remove grille, drain the power steering reservoir and disconnect the oil lines from valve housing. To facilitate reinstallation, mark the relative position of the valve housing with respect to the worm housing with a scribed line. Unbolt and remove the valve housing assembly from tractor.

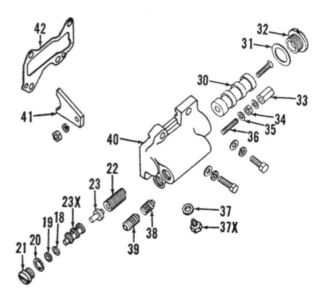

Fig. JD1019 — Exploded view of the steering valve housing used on late models. On earlier models, the valve housing is similar except the system relief valve is mounted within the pump housing as shown in Fig. JD1014.

18. Washers
19. Shim washers
20. Gasket
21. Plug
22. Spring
23. Relief valve
23x. Relief valve sleeve
30. Steering valve
31. Gasket
32. Plug
33. Cap nut
34. Jam nut
35. Washer
36. Adjusting screw
37. Copper washer
37x. Stop screw
38. Flow control valve spring
39. Flow control valve
41. Valve arm

25. To install the valve housing, use a new gasket and tighten the assembly cap screws finger tight; also, be sure to align the previously affixed scribed lines. Reconnect the oil lines, fill the reservoir and start engine. Center the valve housing as in paragraph 23.

26. OVERHAUL. With the valve housing removed, clean the unit and remove the flow control valve adjusting screw. Remove the union adapter from rear of housing and slide the flow control valve and spring out of the valve bore. Remove the end plug from housing and remove the steering valve arm by removing its retaining bolt as shown in Fig. JD1018. On late models, remove the relief valve plug (21—Fig. JD1019) and withdraw the relief valve sleeve, valve and spring.

Wash all parts in a suitable solvent and renew any that are nicked, grooved or worn.

Install the flow control valve spring, valve and adjusting screw, and initially adjust the valve as follows: Using a punch or similar tool, hold the valve against rear of flow valve passage and turn the adjusting screw in until it just contacts the spring; then turn the screw in two additional turns. Install the union adapter, flow valve and arm and end plug. On late models, assemble the relief valve, using the same number of adjusting washers as were removed.

With the valve housing installed, center the unit as in paragraph 23 and adjust the flow control valve as outlined in paragraph 22. On late models, check the system operating pressure as outlined in paragraph 19.

STEERING WORM AND VALVE ACTUATING SLEEVE

27. END PLAY ADJUSTMENT. Two end play adjustments are provided in the steering worm and valve actuating sleeve assembly. The valve actuating sleeve set screw must be adjusted to provide an end play of 0.001-0.002 between the cone point of the set screw and the helix slot in the steering worm shaft. Then, the number of shims between the worm housing and the front bearing housing should be varied to obtain a worm shaft end play of 0.001-0.004. The two adjustments will therefore provide a total end play for the worm and actuating sleeve assembly of 0.002-0.006.

The two adjustments can be approximated, but since the end play should be held as near as possible to the lower limit, it is recommended that a dial indicator be used as follows:

Fig. JD1017 — Adjusting the power steering flow control valve. To make the adjustment, turn the screw in to obtain the fastest turning speed without causing an objectionable amount of front wheel flutter.

Fig. JD1018 — Removing bolt retaining the steering valve arm to the steering valve.

Remove grille and mount a dial indicator with contact button resting on rear end of the valve actuating sleeve as shown in Fig. JD1020. Remove the worm shaft front bearing housing and all of the adjusting shims. Reinstall the front bearing housing without the shims so that the worm shaft bearings have no end play. Remove the inspection hole plug and turn the steering wheel until the actuating sleeve set screw lines up with the inspection hole. Turn the set screw either way as required to obtain an end play of 0.001-0.002. It is desirable, however, to hold the end play adjustment as close to 0.001 as possible. Some early models did not have an inspection hole for adjusting the set screw. Therefore, on these early models, it is necessary to remove the valve housing to make the set screw adjustment.

Note: Adjustment of the set screw will be made easier by grinding a screwdriver tip to the dimensions shown in Fig. JD1021.

With the set screw adjusted, remove the worm shaft front bearing housing, install the previously removed shims

and check the end play which should now be 0.002-0.006. Note: This 0.002-0.006 end play includes the 0.001-0.002 end play previously obtained by the actuating sleeve set screw adjustment. The number of shims under the front bearing housing should be varied to reduce the total end play as near to 0.002 as possible. Shims are available in thicknesses of 0.0025, 0.010 and 0.018.

28. STEERING SHAFT BIND. To obtain proper steering action, there should be a minimum of binding in the steering shaft. To check for binding with engine stopped, turn the steering wheel in either direction without turning the front wheels. When the steering wheel is released, spring pressure against the cam within the valve actuating sleeve should return the steering wheel to a neutral position.

If binding exists after the end play is adjusted as outlined in paragraph 27, check for misalignment of the steering shaft rear support. The support can be aligned after loosening the two retaining cap screws.

An improperly located pedestal can also cause binding. Beginning with the front center pedestal retaining

cap screw, loosen every other cap screw and bump pedestal either way as shown in Fig. JD1022 until the steering shaft is free.

29. BACKLASH. Backlash between the steering worm and gear is controlled by shims located between the worm shaft housing and the pedestal. To check the backlash, rotate the steering wheel back and forth without permitting the steering cams within the actuating sleeve to separate or the front wheels to turn. Backlash should be ⅜-¾ inch when measured at outer edge of steering wheel. Check the backlash with the front wheels in the extreme right, straight ahead and extreme left positions. Excessive variation in backlash between the three positions indicates a bent steering spindle.

29A. R&R AND OVERHAUL. To remove the steering worm and valve actuating sleeve, first remove the grille, disconnect the steering shaft coupling and drain the power steering oil reservoir. Disconnect the oil lines and remove the steering valve housing as outlined in paragraph 24. Unbolt and remove the worm shaft housing from pedestal, being careful not to

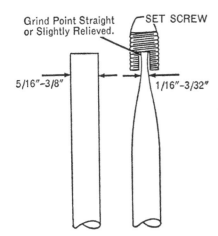

Fig. JD 1020—Adjusting the steering valve actuating sleeve set screw. Adjustment is best accomplished by grinding a screwdriver tip as shown in Fig. JD1021.

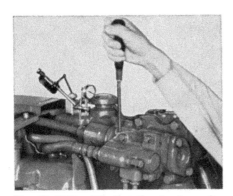

Grind Point Straight or Slightly Relieved.

SET SCREW

5/16"-3/8"

1/16"-3/32"

Fig. JD1021—Dimensions for grinding a special screwdriver for adjusting the valve actuating sleeve set screw.

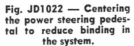

Fig. JD1022 — Centering the power steering pedestal to reduce binding in the system.

Fig. JD1023 — Exploded view of the power steering worm, valve actuating sleeve and associated parts. Backlash between the worm and steering gear is controlled by shims (65).

43. Woodruff key
44. Steering shaft
45. Oil seal
46. Oil seal housing
47. Gasket
48. Valve actuating sleeve
49. Pins
50. Expansion plug
51. Cone point set screw
52. Bearing housing
53. "O" ring
54. Shims
55. Bearing cup
56. Bearing cone
57. Steering worm
58. Worm cam
59. Cam rod
60. Cotter pin
61. Spring
62. Washer
63. Nut
64. Worm housing
65. Shims
66. Pipe plug
67. Dowel pin

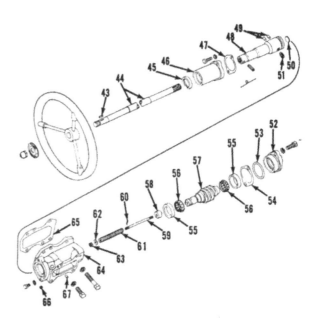

damage or lose the backlash adjusting shims.

Remove oil seal housing (46—Fig. JD1023), actuating sleeve set screw (51) and actuating sleeve (48). When withdrawing the actuating sleeve, it may be necessary to insert a punch through the sleeve hole as shown in Fig. JD1024 to position the cam rod spring so it will clear the internal pins in the sleeve. Unbolt the worm shaft front bearing housing, save the adjusting shims and withdraw the worm. Remove cam spring (61—Fig. JD1023) and cam (58) from rod.

Clean and inspect all parts and renew any that are damaged and cannot be reconditioned. Lip of oil seal (45) goes toward worm. Renew the worm shaft if it is scored or worn at cam end, helix slot or worm. Burrs can be removed from the worm shaft keyway or the surface over which the actuating sleeve operates by using a fine stone. Check the valve actuating sleeve pins for damage, and bore of sleeve for being scored. Check fit of sleeve on worm shaft. If sleeve is tight on shaft or if bore is scored, remove the dowel pins and hone the sleeve to insure a free fit. Renew the pins if they are worn or damaged. Renew the actuating sleeve set screw if it has a loose fit in sleeve or if cone point is worn.

When reassembling, use a valve spring tester and measure the length of spring (61) at 70-80 lbs. pressure.

Note: If spring will not exert 70-80 lbs. pressure at 2⅞ inches or more, obtain a new spring and measure its length at 70-80 lbs. pressure.

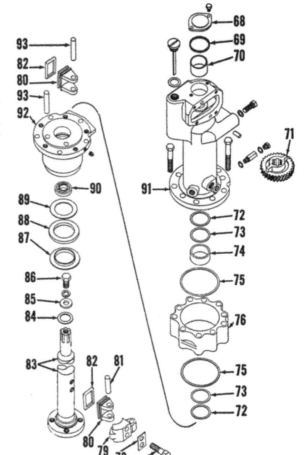

Fig. JD1026 — Exploded view of power steering pedestal, steering spindle and associated cylinder parts.

68. Pedestal cover or cup plug
69. Gasket
70. Bushing
71. Steering gear
72. Back-up gasket
73. "O" ring
74. Bushing
75. Gasket
76. Steering cylinder
77. Special cap screws
78. Lock plate
79. Vane bracket
80. Steering vane
81. Vane pin
82. Vane packing
83. Steering spindle
84. Shim washers
85. Washer
86. Cap screw
87. Retainer
88. Cork seal
89. Washer
90. Thrust bearing
91. Pedestal
92. Spindle quill
93. Vane pin

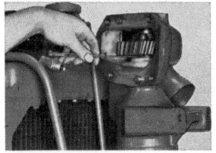

Fig. JD1025—Checking the power steering spindle end play. Desired end play of 0.004-0.021 is controlled by shims under the gear.

Then install the spring on the cam rod and tighten the adjusting nut until spring is compressed to the same length.

Fig. JD1024 — When removing the valve actuating sleeve, it is often necessary to use a punch and position the cam rod spring so it will clear the sleeve pins.

Reassemble the worm and actuating sleeve and adjust the end play as in paragraph 27. Install the unit and adjust the backlash as in paragraph 29. Install the valve housing as outlined in paragraph 25..

SPINDLE, PEDESTAL AND CYLINDER
29B. **R&R AND OVERHAUL.** Remove the grille, disconnect the steering shaft coupling and drain the power steering reservoir. Disconnect the

oil lines, unbolt worm housing from pedestal and remove the worm housing and valve housing assembly being careful not to lose the backlash adjusting shims located between the worm housing and pedestal. Using a feeler gage as shown in Fig. JD1025, measure the spindle end play clearance between the steering gear and the adjusting shims. If the clearance is not within the limits of 0.004-0.021, it will be necessary to add or remove the necessary amount of shims during reassembly. Remove the cup plug (or cover) from top of pedestal and the cap screw and washer retaining the steering gear to vertical spindle. Using a wood block or brass drift and a heavy hammer, drive the steering gear upward until gear is loose on the spindle splines. Unbolt pedestal from front end support and lift pedestal, gear and adjusting shims from spindle.

Remove cylinder (76—Fig. JD1026) and unbolt wheel fork, lower spindle or center steering arm from the steering spindle (83). Unlock the cap screws (77) and unbolt the vane bracket from spindle. Lift the quill (92) and steering spindle assembly from tractor and bump spindle out of quill.

Lower surface of pedestal and upper surface of quill (92) form part of the cylinder sealing surface and must be smooth and free from nicks and scratches. Original inside diameter of the pedestal upper bushing (70) is 1.502-1.503. When installing the bushing, the beveled edge should be toward bottom of pedestal, the split should be toward the steering worm and top of bushing should be flush with top of bushing bore. New bushing is pre-sized and, if not damaged during installation, will require no final sizing. Original inside diameter of the pedestal lower bushing (74) is 2.502-2.503. To remove the bushing, extract the "O" ring located just above the bushing and use a suitable puller. Using a closely fitting mandrel, install bushing until it is flush with bottom of pedestal. Bushing is not pre-sized and must be honed to an inside diameter of 2.502-2.503. After honing, clean and lubricate the bushing. Install a new "O" ring and back-up washer above the lower bushing. Note: The upper end of the bushing forms the lower edge of the "O" ring groove. Install a new and well lubricated neoprane gasket in groove in lower flange of pedestal.

Bearing surface of spindle must be smooth and free from nicks or

Fig. JD1027—Correct installation of the steering vane bracket and vane. Screws (77) should be tightened to a torque of 208 Ft.-Lbs.

scratches. Side of thrust bearing (90) marked "THRUST HERE" must be installed toward bottom of quill (92). Install washer (89) new cork seal (88) and retainer (87). Install back-up washer (72) in quill, then install a new "O" ring on top of the washer With "O" ring (73) well lubricated, slide spindle into quill, position the assembly on the front end support and bolt wheel fork, lower spindle or center steering arm in position and tighten the cap screws to a torque of 275 Ft.-Lbs.

Install the steering vane bracket (79), tighten the retaining cap screws to a torque of 208 Ft.-Lbs. and bend tab of lock plate over cap screw heads. Install vane seals, vanes and pins. Refer to Fig. JD1027. Install cylinder (76—Fig. JD1026) and position the cylinder so that all cap screw holes in cylinder and quill are aligned. Lubricate "O" ring (73) in lower part

of pedestal and install pedestal. When lowering pedestal into position, install the end play adjusting shims (84) and steering gear. Tighten the gear retaining cap screw to a torque of 190 Ft.-Lbs.

Note: The number of end play adjusting shims to be inserted between steering gear and pedestal were determined during disassembly and should provide the steering spindle with an end play of 0.004-0.021 when the pedestal retaining cap screws are securely tightened.

Beginning with the front center cap screw hole in pedestal, install a long cap screw in every other hole. Install the remaining shorter cap screws and tighten all of the pedestal cap screws to a torque of 150 Ft.-Lbs.

Assemble the remaining parts and adjust the steering gear backlash as in paragraph 29. Remove any binding in the steering shaft as outlined in paragraph 28.

ENGINE AND COMPONENTS

The engine crankcase and the tractor main case is an integral unit. A wall in the main case separates the crankcase compartment from the transmission and differential compartment.

CYLINDER HEAD

Models 50-60

30. **REMOVE AND REINSTALL.** To remove the cylinder head, first drain cooling system, loosen hose clamps and remove the water inlet casting (lower water pipe) from cylinder head. Disconnect choke and throttle rod from carburetor. On early models not equipped with the automatic fuel shut-off valve, shut-off the fuel. Disconnect fuel line and remove carburetor, exhaust pipe and manifold. Note: On model 50 all-fuel tractors prior to Ser. No. 5015951 and model 60 all-fuel

tractors prior to Ser. No. 6013900, the manifold is retained by studs and cannot be removed until cylinder head is off.

Remove tool box, ventilator tube and tappet lever cover. Disconnect the tappet lever oil line, remove tappet levers assembly and withdraw push rods. Remove the cylinder head retaining stud nuts, slide cylinder head forward on studs and withdraw head from tractor.

When reassembling, soak cylinder head gasket in engine oil for five seconds, allow to drain for one minute and install gasket with smooth side toward cylinder block. Slide cylinder head on studs and using shellac on mating surfaces of cylinder head and tappet lever oil lead casting, install oil casting on stud (2 or A—Fig.

JD1031). Using new lead washers under the stud nuts, install the nuts and tighten in the numerical sequence shown in Fig. JD1031 if ten studs are used or in the alphabetical sequence if nine studs are used. Tighten the stud nuts to a torque of 104 Ft.-Lbs. for model 50, 150 Ft.-Lbs. for model 60. Reinstall tappet lever assembly and adjust tappets to 0.022 cold.

After engine is hot, re-tighten the head nuts to the specified torque value of 104 Ft.-Lbs. for model 50, 150 Ft.-Lbs. for model 60 and adjust the tappet gap to 0.020 hot.

Model 70

31. **REMOVE AND REINSTALL.** To remove the cylinder head, first drain cooling system, unbolt the water inlet casting from head and remove casting and lower water pipe. Disconnect

choke and throttle rods from carburetor. On early models not equipped with the automatic fuel shut-off valve, shut-off the fuel. Disconnect fuel line and remove carburetor and lower half of air intake elbow. Remove exhaust pipe, unbolt and remove manifold. Remove ventilator tube and tappet lever cover. Disconnect the tappet lever oil line, remove tappet levers assembly and withdraw push rods. Remove the front two implement mounting cap screws from front end support. Remove the cylinder head retaining stud nuts, slide cylinder head forward on studs and withdraw head from tractor.

When reassembling, soak cylinder head gasket in engine oil for five seconds, allow to drain for one minute and install gasket with smooth side toward cylinder block. Slide cylinder head on studs and using shellac on mating surfaces of cylinder head and tappet lever oil lead casting, install oil lead casting on stud (2 or A—Fig. JD1031). Using new lead washers under the stud nuts, install the nuts and tighten in the alphabetical sequence shown in Fig. JD1031 if nine studs are used, numerical sequence if ten studs are used and to a torque value of 208 Ft.-Lbs. Install tappet levers assembly and adjust tappets to 0.022 cold.

Note: When installing push rods and tappet levers assembly, remove crankcase cover to feel action of cam followers when they are engaged by push rods. Stuff shop towels in push rod holes in cylinder head to hold push rods back against cam followers until tappet levers assembly is bolted down. Make certain that tappet lever supports (7—Fig. JD1033) engage the positioning dowels (12) in the cylinder head.

After engine is hot, re-tighten the head nuts to the specified torque value of 208 Ft.-Lbs. and adjust the tappet gap to 0.020 hot.

VALVES AND SEATS
Models 50-60-70 (Except LP-Gas)

32. Intake and exhaust valves are not interchangeable. Valves seat directly in cylinder head with a seat angle of 30 degrees for the intake and 45 degrees for the exhaust. Valve face angle is as follows:

Intake:

Models 50-60 ....29¾ Degrees
Model 70 All-Fuel .....30 Degrees
Model 70 Gasoline ....29¾ Degrees

Exhaust:

Models 50-60 ....44½ Degrees
Model 70 All-Fuel .....45 Degrees
Model 70 Gasoline ...44½ Degrees

Desired seat width is ⅛-9/64 inch. Seats can be narrowed, using 20 and 70 degree stones.

Intake and exhaust valve stem diameter is as follows:

Model 50 ....0.4335-0.4345
Model 60 ....0.496-0.497
Model 70 All-Fuel 0.558-0.559
Model 70 Gasoline......0.496-0.497

Adjust intake and exhaust tappet gap to 0.020 hot. If gap is adjusted with engine cold, allow 0.002 for expansion purposes and recheck the gap when engine is hot.

Note: On any 50, 60 or 70 tractor on which sticking of exhaust valves occurs, it is recommended that exhaust valve rotators be installed. The necessary rotators and springs are available from John Deere. Refer to paragraph 35.

Models 50-60-70 (LP-Gas)

33. Intake valves are not interchangeable with the exhaust valves which are factory equipped with positive type rotators. Refer to paragraph 35 for valve rotator information. Intake valves seat directly in cylinder head with a face angle of 29¾ degrees and a seat angle of 30 degrees. Exhaust valves seat on renewable, alloy seat inserts with a face angle of 44½ degrees and a seat angle of 45 degrees.

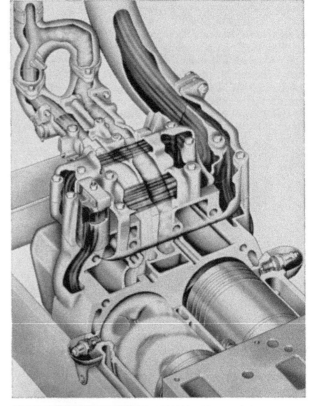

Fig. JD1030 — Cut-away view of model 60 gasoline carburetor, cylinder head, manifold and associated parts. For the purposes of this illustration, the models 50 and 70 are similar. The manifold is equipped with a valve which can be set for hot or cold operation.

Fig. JD1031—Model 50 cylinder head with tappet levers and push rods removed. Models 60 and 70 are similar. If nine studs are used, tighten the stud nut in the alphabetical sequence; if ten studs are used, use the numerical sequence.

Desired seat width is 1/8-9/64 inch. Seats can be narrowed using 20 and 70 degree stones. Refer to paragraph 34 for method of renewing the valve seat inserts. Intake and exhaust valve stem diameter is 0.496-0.497 for models 60 and 70, 0.4335-0.4345 for model 50.

Adjust intake and exhaust tappet gap to 0.020 hot. If gap is adjusted with engine cold, allow 0.002 for expansion purposes and recheck the gap when engine is hot.

34. Exhaust valve seat inserts can be removed by using a suitable puller. New inserts are 0.010 oversize and the head should be reamed to provide a 0.003 interference fit. Thoroughly clean the counterbore in cylinder head before attempting to install a new seat. Chill seat in dry ice to facilitate installation.

After installation, check the concentricity of the seat with respect to the valve guides. Seats should be concentric within 0.002.

VALVE ROTATORS
Models 50-60-70

35. Positive type exhaust valve rotators (Fig. JD1034) are factory installed on all model 60 and 70 LP-Gas tractors and are available for field installation on any other 50, 60 or 70 tractor. The rotators can be considered in satisfactory operating condition if the exhaust valves rotate a slight amount each time the valves open. Servicing consists of renewing the complete rotator unit.

VALVE GUIDES AND SPRINGS
Models 50-60-70

36. Intake and exhaust valve guides are interchangeable in any one model

and can be driven from cylinder head if renewal is required. Press new guides into cylinder head so that smaller O.D. of guides will be toward valve springs. Distance from port end of guides to gasket surface of cylinder head is as follows:

Model 50 .... 1 31/32 inches

Model 60 .... 1 25/32 inches

Model 70 All-Fuel ... 2 1/16 inches

Model 70 Gasoline &
  LP-Gas .... 2 1/4 inches

Ream new guides after installation to the following inside diameter:

Model 50 .... 0.4386-0.4400

Model 60 .... 0.5009-0.5025

Model 70 All-Fuel ... 9.5634-0.5650

Model 70 Gasoline &
  LP-Gas 0.5009-0.5025

Recommended clearance between valve stems and guides is as follows:

Model 50 .... 0.0041-0.0065

Model 60 .... 0.0039-0.0065

Model 70 All-Fuel 0.0044-0.0070

Model 70 Gasoline &
  LP-Gas 0.0039-0.0065

37. Intake and exhaust valve springs are interchangeable in all models except those equipped with valve rotators. Renew any spring which is rusted, discolored or does not meet the load test specifications which follow:

Pounds Test @ Height in Inches:

Model 50 In. ...... 35-39 @ 2 13/16

Model 50 Exh. No
  Rotators .... 35-39 @ 2 13/16

Model 50 Exh. With
  Rotators .... 31.5-38.5 @ 2 19/32

Model 60 In. .... 36-44 @ 2 3/4

Model 60 Exh. No
  Rotators .... 36-44 @ 2 3/4

Model 60 Exh. With
  Rotators .... 36-44 @ 2 17/32

Model 70 In. .... 56-68 @ 3 3/8

Model 70 Exh. No
  Rotators .... 56-68 @ 3 3/8

Model 70 Exh. With
  Rotators .... 45-55 @ 3 7/64

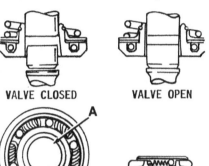

VALVE CLOSED   **VALVE OPEN**

SECTION A-A

Fig. JD1034—Positive type exhaust valve rotators as factory installed on models 60 and 70 LP-Gas tractors. The units are available for field installation on any 50, 60 or 70 tractor. Servicing of the rotator consists of renewing the complete unit.

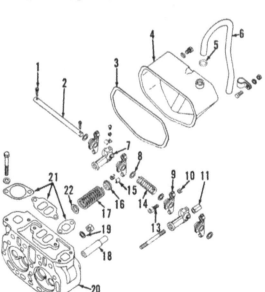

1. Cotter pin
2. Tappet lever shaft
3. Gasket
4. Tappet lever cover
5. "O" ring
6. Vent tube
7. Support
8. Washer
9. Tappet lever
10. Lock nut
11. Nut
12. Dowels (Model 70)
13. Tappet adjusting screw
14. Spring
15. Keepers
16. Valve spring cap
17. Valve spring
18. Valve guide
19. Lead washer
20. Cylinder head
21. Manifold gaskets
22. Valve spring centering washer (Model 50)

Fig. JD1032—Exploded view of model 50 dual induction cylinder head, tappet levers and associated parts.

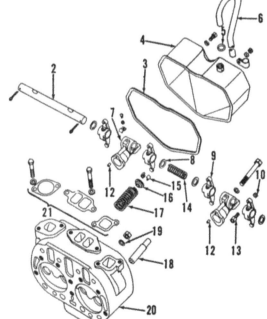

Fig. JD1033—Exploded view of model 70 cylinder head and associated parts. Model 60 construction is similar except for details of support brackets (7). LP-Gas cylinder heads are fitted with exhaust valve seat inserts.

VALVE TAPPET LEVERS
(ROCKER ARMS)
Models 50-60-70

38. **R&R AND OVERHAUL.** The rocker arms assembly can be removed after removing the tappet lever cover and disconnecting the tappet lever oil line. Check the tappet levers and shaft against the values which follow:

Shaft Diameter:
Model 50 . . . . 0.606-0.607
Model 60 . . . . 0.984-0.985
Model 70 . . . . 1.234-1.235

Tappet Lever Bore:
Model 50 . . . . 0.610-0.612
Model 60 . . . . 0.986-0.991
Model 70 . . . . 1.236-1.238

Shaft Clearance In Lever:
Model 50 . . . . 0.003-0.006
Model 60 . . . . 0.001-0.007
Model 70 . . . . 0.001-0.004

Excessive wear of any of the component parts of the tappet lever assembly is corrected by renewing the parts. Intake and exhaust tappet levers are interchangeable in any one model.

When reinstalling, adjust tappet gap to 0.020 hot.

NOTE: When installing push rods and tappet levers assembly on model 70, remove crankcase cover to feel action of cam followers when they are

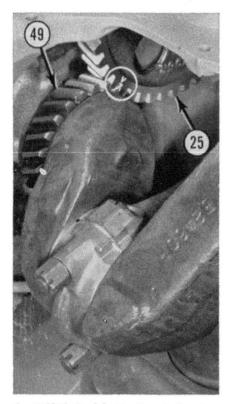

Fig. JD1035—Models 50, 60 and 70 valves are properly timed when "V" marked tooth space on camshaft gear (25) is meshed with "V" marked tooth on crankshaft gear (49).

engaged by push rods. Stuff shop towels in push rod holes in cylinder head to hold push rods back against cam followers until tappet levers assembly is bolted down. Make certain that tappet lever supports (7—Fig. JD1033) engage the positioning dowels (12) in the cylinder head.

VALVE TIMING
Models 50-60-70

39. Valves are properly timed when "V" marked tooth space on camshaft gear is in register with the "V" marked tooth on the crankshaft gear as shown in Fig. JD1035. The gear marks can be viewed after removing the crankcase cover.

Valve timing, however, can be checked without removing crankcase cover, as follows:

First make certain that valve tappet gap is adjusted to 0.020 hot. Turn crankshaft until the exhaust valve of No. 1 (LEFT) cylinder is just beginning to open (0.000 tappet gap). At

Fig. JD1036—Models 50, 60 and 70 valve timing mark "LHEO" is located on the outer rim of the flywheel. The mark can be viewed through the flywheel housing inspection port as shown.

this time, the flywheel mark "LHEO" should be approximately in line with notch in the timing hole in flywheel cover as shown in Fig. JD1036.

TIMING GEARS
Models 50-60-70

40. The procedure for renewing the camshaft gear and/or crankshaft gear is outlined in the paragraphs which cover the renewal of the respective shafts.

CAMSHAFT AND BEARINGS
Model 50

45. **R&R AND OVERHAUL.** To remove the camshaft gear and/or shaft and bearings, first remove governor and housing assembly as outlined in paragraph 102, clutch and belt pulley as in paragraph 118 and the reduction gear cover as outlined in paragraph 123. Remove the crankcase cover, tappet lever cover, rocker arms assembly and push rods. Working on inside of crankcase, disconnect oil line and remove the oil line connector (35—Fig. JD1037) from the camshaft right bearing (34). Remove the camshaft right and left bearings and completely unscrew the three cap screws (A—Figs. JD1037 & 1038) which retain the cam follower guide to the main case.

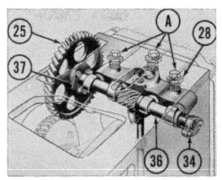

Fig. JD1038—Model 50 camshaft installation. The cam follower guide is retained by cap screws (A).

Fig. JD1037 — Exploded view of model 50 camshaft, gear and followers. On late production tractors, cap screws (33) are locked in place by tab washers.

A. Cap screws
23. Left bearing
24. Shim gaskets
25. Cam gear
26. Cam followers
27. Cam follower guide
28. Guide support
29. Push rods
30. Oil pump drive gear
31. Gasket
32. Drive gear bearing
34. Right bearing
35. Oil line connector
36. Camshaft
37. Locking plate

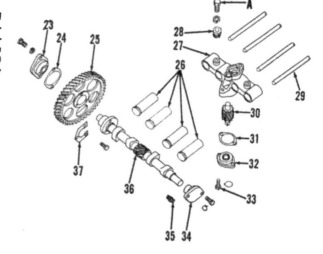

Straighten tabs on locking plate (37—Fig. JD1037) and unbolt gear from shaft. Buck up the camshaft gear and bump camshaft toward right and out of gear. The camshaft gear can be removed from above at this time.

With the camshaft gear removed, disconnect both of the oil lines which connect to top of inside of crankcase. Move camshaft toward right and withdraw shaft through crankcase cover opening as shown in Fig. JD1045.

The 0.7505-0.7515 diameter camshaft bearing journals have a normal operating clearance of 0.0015-0.0045 in the 0.753-0.755 diameter cast iron bearings. If the running clearance is excessive, renew the shaft and/or bearings.

When reassembling, chamfer (C—Fig. JD1039) on camshaft gear must be toward shoulder on camshaft. Install gear on camshaft so that when right hand lobe (L) is pointing upward, the "V" mark on gear is at the top. Bolt gear to camshaft and lock the nut and cap screws in position by upsetting tabs of lock plate (37—Fig. JD1037).

Mesh camshaft gear with crankshaft gear so that timing marks are in register as shown in Fig. JD1040 and vary the number of shims (24—Fig. JD1037) which are located between the left bearing and the main case to give the camshaft an end play 0.015-0.030 when checked with a dial indicator as shown in Fig. JD1041. Install the remaining parts by reversing the disassembly procedure, check the ignition timing as in paragraph 113 and adjust the valve tappet gap to 0.020 hot.

Fig. JD1041—Using a dial indicator to check model 50 camshaft end play. Desired end play of 0.015-0.030 is controlled by shims (24) located under bearing (23).

A. Cap screws
23. Left bearing
25. Cam gear
28. Guide support

Model 60

46. **R&R AND OVERHAUL.** To remove the camshaft gear and/or shaft and bearings, first remove the flywheel as outlined in paragraph 59, governor as in paragraph 103, clutch and belt pulley as in paragraph 118 and the reduction gear cover as outlined in paragraph 130. Remove the camshaft left bearing housing (38—Fig. JD1042) and using a suitable puller as shown in Fig. JD1043, remove the shaft left bearing cone. Remove the crankcase cover, tappet lever cover, rocker arms assembly and push rods. Completely unscrew the three cap screws (A—Figs. JD1042 & 1044) which retain the cam follower guide to the main case. Straighten tabs on locking plate (37—Fig. JD1042), and unbolt gear from shaft. Buck up the camshaft gear and bump camshaft toward right and out of gear. The camshaft gear can be removed from above at this time.

With the camshaft gear removed, disconnect both of the oil lines which connect to top of inside of crankcase. Move camshaft toward right and withdraw shaft through crankcase cover opening as shown in Fig. JD1045.

Fig. JD1043—Using a special puller to remove the model 60 camshaft left bearing cone. The procedure on model 70 is similiar.

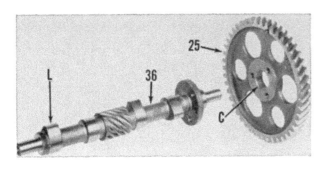

Fig. JD1039 — When installing camshaft gear on models 50, 60 and 70, chamfer (C) must be toward shoulder on shaft and right lobe (L) must point in same direction as (V) mark on gear.

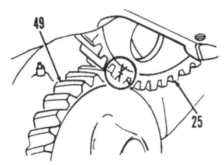

Fig. JD1040—Models 50, 60 and 70 valves are properly timed when "V" marked tooth space on camshaft gear (25) is meshed with "V" marked tooth on crankshaft gear (49).

Fig. JD1042 — Exploded view of model 60 camshaft and associated parts. Camshaft end play is controlled by spring (42).

A. Cap screw
25. Cam gear
26. Cam followers
27. Cam follower guide
28. Guide support
29. Push rods
30. Oil pump drive gear
31. Gasket
32. Drive gear bearing
33. Cap screw
36. Camshaft
37. Locking plate
38. Left bearing housing
39. Gasket
40. Bearing cup
41. Bearing cone
42. Spring

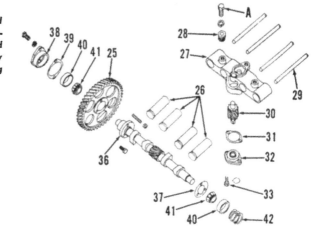

When reassembling, chamfer (C—Fig. JD1039) on camshaft gear must be toward shoulder on camshaft. Install gear on camshaft so that when right hand lobe (L) is pointing upward, the "V" mark on gear is at the top. Bolt gear to camshaft and lock the nut and cap screws in place by upsetting tabs of lock plate (37 — Fig. JD1042). Mesh camshaft gear with crankshaft gear so that timing marks are in register as shown in Fig. JD1040 and install the left bearing housing. Camshaft end play is controlled by a spring at right end of shaft. Make certain that the spring is in position when installing the reduction gear cover. Install the remaining parts by reversing the disassembly procedure, check the ignition timing as outlined in paragraph 113 and adjust the valve tappet gap to 0.020 hot.

Model 70

47. **R&R AND OVERHAUL.** To remove the camshaft gear and/or shaft and bearings, first remove the flywheel as outlined in paragraph 59, governor as in paragraph 103, clutch and belt pulley as in paragraph 119 and the reduction gear cover as outlined in paragraph 136. Remove the camshaft left bearing housing (38—Fig. JD1046) and using a suitable puller as shown in Fig. JD1043, remove the shaft left bearing cone. Remove the crankcase cover, tappet lever cover, rocker arms assembly and push rods. Remove the first reduction gear, and on models so equipped, remove the powershaft idler gear. On models with live power take-off, remove snap ring and using a suitable puller, remove the power shaft drive gear as shown in Fig. JD1047. Unwire and remove cap screws retaining the right main bearing housing to main case and remove the housing. Support crankshaft and remove oil slinger cover, slinger and left main bearing housing. Lower crankshaft, with rods attached, down enough to clear camshaft. Straighten tabs on locking plate (37—Fig. JD1046) and unbolt gear from shaft. Buck up the camshaft gear and bump the camshaft toward right and out of gear. The camshaft gear can be removed from above at this time.

With the camshaft gear removed, disconnect both of the oil lines which connect to top of inside of crankcase. Move camshaft toward right and withdraw shaft through crankcase cover opening as shown in Fig. JD1045.

When reassembling, chamfer (C—Fig. JD1039) on camshaft gear must be toward shoulder on camshaft. Install gear on camshaft so that when right hand lobe (L) is pointing upward, the "V" mark on gear is at the top. Bolt gear to camshaft and lock the nut and cap screws in place by upsetting tabs of lock plate (37 — Fig. JD1046). Raise the crankshaft into position and assemble the main bearing housings. Mesh camshaft gear with crankshaft gear so that timing marks are in register as shown in Fig. JD1040 and install the camshaft left bearing housing. Using a brass drift, install the powershaft drive gear as shown in Fig. JD1048. Install the drive gear re-

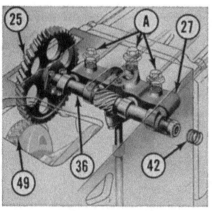

Fig. JD1044—Model 60 camshaft installation. Camshaft end play is controlled by spring (42).

Fig. JD1046 — Exploded view of model 70 camshaft and associated parts. Cam followers (26) operate directly in the main case bores.

25. Cam gear
26. Cam followers
29. Push rods
30. Oil pump drive gear
31. Gasket
32. Drive gear bearing
33. Cap screw
36. Camshaft
37. Locking plate
38. Left bearing housing
39. Gasket
40. Bearing cup
41. Bearing cone
42. Spring
43. Bushing

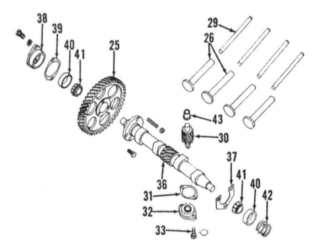

Fig. JD1047—Using puller to remove the continuous power shaft drive gear on model 50 tractor. The same procedure can be used on models 60 and 70.

Fig. JD1045—Removing model 60 camshaft through crankcase opening. A similar procedure is used on model 50.

taining snap ring, powershaft idler gear and first reduction gear.

Camshaft end play is controlled by a spring at right end of shaft. Make certain that spring is in position when installing the reduction gear cover. Leave crankcase cover off until after push rods and rocker arms have been installed so as to feel action of cam followers when they are engaged by the push rods. Stuff shop towels in push rod holes in cylinder head to hold push rods back against cam followers until after rocker arms assembly is bolted down. Make certain that tappet lever supports engage the positioning dowels in the cylinder head. Install the remaining parts by reversing the removal procedure, check ignition timing as outlined in paragraph 113 and adjust the valve tappet gap to 0.020 hot.

CAM FOLLOWERS
Models 50-60

48. The cylindrical type cam followers (26—Figs JD1037 or 1042) operate directly in the unbushed bores of follower guide (27). To remove the followers and guide assembly, first remove camshaft as outlined in paragraph 45 or 46. Remove the oil pump cover and idler gear. Grasp drive gear and pull drive gear and shaft down and out of pump body. Refer to Fig. JD1049. Withdraw follower guide and followers from above.

Excessive wear between the cam followers and the guide is corrected by renewal of the component parts.

Model 70

49. The mushroom type cam followers (26—Fig. JD1046) operate directly in machined bores in the tractor main case. To remove the followers, remove

camshaft as outlined in paragraph 47 and withdraw followers from the case bores. Excessive clearance between the followers and the case bores is corrected by renewal of the parts.

ROD AND PISTON UNITS
Models 50-60-70

50. To remove the connecting rod and piston units, first remove cylinder head as outlined in paragraphs 30 or 31. Remove platform, hydraulic lines and the "Powr-Trol" pump. Remove the crankcase cover, connecting rod caps, bearing inserts and rod bolts. Remove carbon accumulation and ridge from unworn portion of cylinders to prevent damage to ring lands and to facilitate piston removal.

Using a piece of 2x4 wood, make up a set of blocks shown in Fig. JD1051. Using the smallest block between connecting rod and crankshaft, turn crankshaft and push rod and piston unit forward. Continue this process with the next largest block, and so on, until pistons can be withdrawn from front.

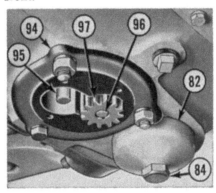

Fig. JD1049—Bottom view of models 50, 60 and 70 main case, showing the oil pump and filter installation. The oil pump cover and idler gear have been removed.

When reinstalling, numbers on rod and cap should be in register and face toward top of engine. Number one cylinder is on left side of tractor. Tighten rod bolts to a torque 85 Ft.-Lbs. for models 50 and 70 and 105 Ft.-Lbs. for model 60.

PISTONS AND RINGS
Models 50-60-70

51. Pistons and rings are available in standard as well as 0.045 oversize for all models. Rings should be installed with dot toward head (closed

Fig. JD1050—Models 50 and 60 connecting rod and piston units. Model 70 is similar except four bolts are used on each rod. Rod bearings are of the precision insert type. Pins are of the full floating type.

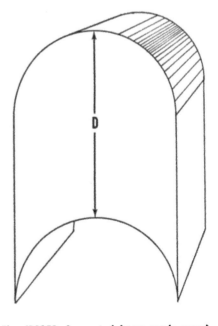

Fig. JD1051—Suggested home made wood block which can be used to push models 50, 60 and 70 connecting rod and piston units forward, out of cylinder block. It is more convenient to use three blocks. Dimension (D) should be approximately as follows: 1st block, 2⅞ inches; 2nd block, 4 inches; third block 5⅝ inches. Refer to text.

Fig. JD1048 — Using a hammer and brass drift to install the continuous power shaft drive gear on model 50. The same procedure can be used on models 60 and 70.

end) of piston. On some models, the compression ring nearest closed end of piston is chrome plated.

Inspect pistons, rings and cylinder walls using the values which follow:

Standard Cylinder Bore

Model 50 ....4.6885-4.6901
Model 60 ....5.4996-5.501
Model 70 All-Fuel ...6.1245-6.1259
Model 70 Gasoline &
  LP-Gas 5.8745-5.8759

Piston Skirt Diameter

Model 50 ....4.6814-4.6830
Model 60 ....5.4930-5.4945
Model 70 All-Fuel 6.1170-6.1185
Model 70 Gasoline &
  LP-Gas 5.8665-5.8675

Piston Skirt Clearance In Bore

Model 50 ....0.0055-0.0087
Model 60 ....0.0051-0.0080
Model 70 All-Fuel ...0.0060-0.0089
Model 70 Gasoline &
  LP-Gas 0.0070-0.0094

Piston Ring Side Clearance

Model 50 (Comp. Rings).0.003-0.006
Model 50 (Oil Rings) ...0.001-0.004
Model 60 (1st & 2nd
  Comp. Rings) 0.002-0.0045
Model 60 (3rd & 4th
  Comp. Rings) 0.001-0.0035
Model 60 (Oil Rings) ..0.001-0.0036
Model 70 Gasoline &
  LP-Gas (Comp. Rings) 0.004-0.006
Model 70 Gasoline &
  LP-Gas (Oil Rings) .0.0015-0.0035
Model 70 All-Fuel (1st
  Comp. Ring) 0.002-0.005
Model 70 All-Fuel (2nd,
  3rd & 4th Comp. Rings
  and Oil Rings) 0.001-0.004

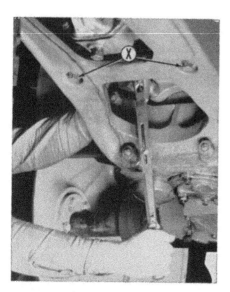

Fig. JD1052 — Tightening stud nuts which retain models 50, 60 and 70 cylinder block to the main case. Cap screws (X) secure block to front end support.

PISTON PINS AND BUSHINGS
Models 50-60-70

52. The full floating type piston pins are retained in the piston pin bosses by snap rings and are available in standard as well as oversizes of 0.003 (marked yellow) and 0.005 (marked red).

Piston pin diameter is as follows:
Model 50 ....1.4165-1.4170
Model 60 ....1.7495-1.7500
Model 70 ....1.7496-1.7500

Fit piston pins to a thumb press fit in piston and to the following clearance in the connecting rod bushing.
Models 50-60 ....0.001-0.0025
Model 70 Gasoline and
  All-Fuel 0.001-0.0024
Model 70 LP-Gas 0.001-0.0034

CONNECTING RODS AND BEARINGS
Models 50-60-70

53. Connecting rod bearings are of the steel-backed, babbitt-lined, slip-in, precision type which can be renewed after removing crankcase cover and connecting rod caps. Although not absolutely necessary, it is more convenient and time will be saved by removing platform, hydraulic lines and the "Powr-Trol" pump.

When installing new bearing shells, make certain that the bearing shell projections engage the milled slot in connecting rod and bearing cap and that cylinder numbers on rod and cap are in register and face toward top of engine. Number one cylinder is on left side of tractor. Connecting rod bearing inserts are available in standard as well as undersizes of 0.002, 0.004, 0.020 and 0.022.

Check the crankshaft and bearing inserts against the values which follow:

Crankshaft Crankpin Diameter
Model 50 ....2.7484-2.7500
Model 60 ....2.9985-3.000
Model 70 ....3.3736-3.3750
Rod Bearing Running Clearance
Models 50-60-70 ....0.001-0.004
Rod Bolt Tightening Torque
Models 50-70 ....85 Ft.-Lbs.
Model 60 ....105 Ft.-Lbs.

CYLINDER BLOCK
Models 50-60-70

54. **REMOVE AND REINSTALL.** To remove the cylinder block, first remove cylinder head as outlined in paragraph 30 or 31 and the connecting rod and piston units as outlined in paragraph 50. Remove spark plug covers, disconnect plug wires and remove spark plugs. Disconnect the heat indicator bulb and remove the upper wa-

ter pipe rear casting (water outlet casting).

On model 70, remove the four implement mounting cap screws from front end support.

On all models, remove cap screws (X—Fig. JD1052) retaining cylinder block to front end support and remove stud nuts retaining block to main case. Slide block forward and withdraw from tractor.

Reinstall cylinder block by reversing the removal procedure and tighten the stud nuts which retain cylinder block to main case to a torque of 167 Ft.-Lbs. for models 50 and 60 and 208 Ft.-Lbs. for model 70.

CRANKSHAFT, SEALS AND MAIN BEARINGS
Models 50-60-70

54A. The crankshaft is carried in two main bearing housings which are fitted with sleeve type bearings. The sleeve-type main bearings are available as individual repair parts and they must be sized after installation in the main bearing housings to provide a shaft diametral clearance of 0.004-0.006 for models 50 and 60, 0.0045-0.0065 for model 70. Main bearings are also available already installed in the main bearing housings on a factory exchange basis. The exchange main bearing and housing units are pre-sized and are available in standard size as well as undersizes of 0.002 and 0.004. Standard size main bearing bore is as follows:
Model 50 (Right) 2.6285-2.6295
Model 50 (Left) 2.2535-2.2545
Model 60 (Right) 3.0040-3.0050
Model 60 (Left) 2.7540-2.7550
Model 70 (Right &
  Left) 3.2545-3.2555

Fig. JD1053 — Checking clearance between model 60 crankshaft and main bearings. The same procedure can be used on models 50 and 70.

Before removing the main bearing housings, check for excessive clearance between crankshaft and main bearings by mounting a dial indicator so that contact button is resting on crankshaft near the main bearing housings. Move crankshaft up and down as shown in Fig. JD1053 and observe the main bearing clearance as shown on the dial indicator. Main bearing running clearance should be 0.004-0.006 for models 50 and 60, 0.0045-0.0065 for model 70.

The right main bearing housing is fitted with two lip type oil seals which should be installed back to back to prevent mixing of transmission oil with the engine crankcase oil. Oil leakage at the left main bearing is prevented by an "O" ring and oil slinger.

55. MAIN BEARINGS. Although most repair jobs associated with the crankshaft and main bearings will require removal of both main bearing housings, there are infrequent instances where the failed or worn part is so located that repair can be accomplished safely by removing only one of the housings. In effecting such localized repairs, time will be saved by observing the following paragraphs as a general guide.

56. RIGHT MAIN BEARING HOUSING AND SEALS. To remove the right main bearing housing, first remove clutch and belt pulley as outlined in paragraph 118 for models 50 and 60 or 119 for model 70. Remove the reduction gear cover as in paragraphs 123, 130 or 136. Withdraw spacer (S—Fig. JD1054) and remove the first reduction gear (RG). On models equipped with engine driven power shaft, remove the power shaft idler gear (IG), remove snap ring (SR) and using a puller as shown in Fig. JD1047, remove the

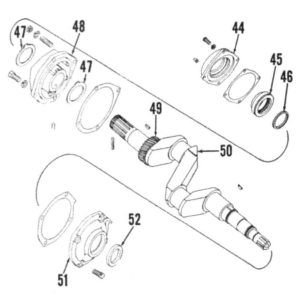

Fig. JD1055 — Exploded view of model 60 crankshaft main bearings and seals. For the purposes of this illustration, models 50 and 70 are similar. The crankshaft in model 70 tractors is counterbalanced. Two oil seals (52) are installed back-to-back.

44. Oil slinger housing
45. Oil slinger
46. "O" ring seal
47. Thrust washer
48. Left main bearing and housing
49. Crankshaft gear
50. Crankshaft
51. Right main bearing and housing
52. Oil seal (two used)

power shaft drive gear. On model 60 tractors prior to Ser. No. 6025000, remove crankcase cover and disconnect oil line from main bearing housing.

On all models, unbolt and withdraw the main bearing housing. The two opposed oil seals (52—Fig. JD1055) can be renewed at this time. When installing new seals, there should be a gap of 1/16 inch between inner seal and main bearing bushing. If seal is pressed in beyond this point, the oil return hole will be restricted. Apply gun grease to lips of seal before installing housing.

Reinstall main bearing housing, using a thin sleeve (or shim stock) to guide oil seals over crankshaft.

NOTE: On model 60 tractors after Ser. No. 6024999, and all models 50 and 70, make certain that oil groove in main bearing housing is positioned over the oil feed hole in the main case. On model 60 tractors prior to Ser. No. 6025000, reconnect oil line to main bearing housing.

Tighten the bearing housing retaining cap screws to a torque of 100-Ft.-Lbs. on model 50, 150 Ft.-Lbs on models 60 and 70 and install safety wire in the drilled cap screw heads. Using a brass drift, install the powershaft drive gear as shown in Fig. JD1048.

57. LEFT MAIN BEARING HOUSING, OIL SLINGER & SEAL. To remove the left main bearing housing, first remove flywheel cover and flywheel as outlined in paragraph 59. Remove the oil slinger housing (44—Fig. JD1055). Mark the relative position of the oil slinger with respect to the crankshaft and remove the oil slinger (45). The rubber seal (46) inside the oil slinger can be renewed at this time.

On model 60 tractors prior to Ser. No. 6025000, remove crankcase cover and disconnect oil line from main bearing housing.

On all models, unbolt and withdraw the left main bearing housing. Renew thrust washers (47) if they are damaged or show wear.

NOTE: On model 60 tractors after Ser. No. 6024999, and all models 50 and 70, make certain that oil groove in main bearing housing is positioned over the oil feed hole in the main case. On model 60 tractors prior to Ser. No. 6025000, reconnect oil line to main bearing housing. Tighten the bearing housing retaining cap screws to a torque of 100 Ft.-Lbs. on model 50, 150 Ft.-Lbs. on models 60 and 70.

When reassembling, make certain that the previously affixed marks on crankshaft and oil slinger are aligned, install the oil slinger housing and tighten the housing retaining cap screws finger tight. Using 0.003 feeler gage or shim stock (SS) as shown in Fig. JD1056, make certain that the oil slinger housing is centered about the slinger and tighten the cap screws.

Fig. JD1054—Right side of model 60 main case with reduction gear cover removed. Models 50 and 70 are similar.

DG. Power shaft drive gear
IG. Power shaft idler gear
RG. First reduction gear
S. Spacer
SR. Snap ring
50. Crankshaft

NOTE: If position of oil slinger with respect to crankshaft is questionable, check the following: Observe right side of flywheel near the hub where a small drive pin is located. This pin must engage the slot in the oil slinger when "V" mark on left side of flywheel and "V" mark on end of crankshaft are in register.

When reassembling, adjust the crankshaft end play as outlined in paragraph 60.

58. CRANKSHAFT. To remove the crankshaft, remove platform, hydraulic lines and "Powr-Trol" pump. Remove crankcase cover, connecting rod caps and rod bearing inserts. Turn crankshaft and push the connecting rod and piston units forward. Support crankshaft and remove both main bearing housings as outlined in paragraphs 56 and 57. Withdraw crankshaft from main case being careful not to nick or damage the bearing journals.

Check the crankshaft and bearings against the values which follow:

Main Journal Diameter

Model 50 (Right) 2.6235-2.6245
Model 50 (Left) 2.2485-2.2495
Model 60 (Right) 2.9990-3.0000
Model 60 (Left) 2.7490-2.7500
Model 70 (Right &
    Left) 3.2490-3.2500

Main Bearing Bushings Inside Diameter

Model 50 (Right) 2.6285-2.6295
Model 50 (Left) 2.2535-2.2545

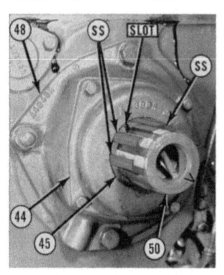

Fig. JD1056—Models 50, 60 and 70 oil slinger housing (44) can be centered about the oil slinger (45) by inserting 0.003 shim stock (SS) between oil slinger and housing before tightening the retaining cap screws. Notice the "V" mark on the end of the crankshaft. This mark must register with a similar mark on the flywheel.

Model 60 (Right) 3.0040-3.0050
Model 60 (Left) 2.7540-2.7550
Model 70 (Right &
    Left) 3.2545-3.2555

Desired Main Bearing Clearance

Models 50-60 ....0.004-0.006
Model 70 ....0.0045-0.0065

Refer to paragraph 53 for crankpin diameter and rod bearing running clearances.

If crankshaft gear is damaged, it can be pulled from crankshaft at this time. Heat new gear in hot oil or water and install on crankshaft so that timing mark on gear is toward crankshaft web (inside of case) and shoulder on gear is toward end of crankshaft (outside of case).

When reassembling, make certain that "V" mark on crankshaft gear is in register with "V" mark on camshaft as shown in Fig. JD1057 and adjust the crankshaft end play as outlined in paragraph 60.

FLYWHEEL AND CRANKSHAFT END PLAY
Models 50-60-70

59. To remove the flywheel, disconnect starter pedal linkage on model 60 or remove starter button on models 50 and 70. Remove flywheel cover and

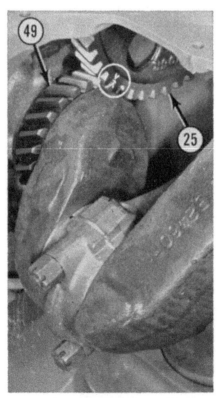

Fig. JD1057—Models 50, 60 and 70 valves are properly timed when "V" marked tooth space on camshaft gear (25) is meshed with "V" marked tooth on crankshaft gear (49).

lock nut and loosen the lock bolts. Bump or pull flywheel from crankshaft.

To install starter ring gear on flywheel, heat gear to approximately 550 degrees F. and place gear on flywheel so that beveled end of ring gear teeth face toward engine crankcase.

When installing flywheel, observe the engine side of the flywheel near the hub where a small drive pin is located. This pin must engage the slot in the crankshaft oil slinger when "V" mark on left side of flywheel and "V" mark on end of crankshaft are in register. Adjust the crankshaft end play as follows:

60. The crankshaft end play is controlled by the position of the flywheel on the crankshaft. To adjust the end play, proceed as follows: Drive flywheel on crankshaft and mount a dial indicator so that indicator contact button is resting on flywheel as shown in Fig. JD1058. Engage clutch, move clutch lever back and forth and observe crankshaft end play as shown in the dial indicator. Continue driving flywheel on crankshaft until the desired end play of 0.005-0.010 is obtained. Tighten both flywheel clamp bolts securely and install the lock nut. Secure lock nut in position by peening a portion of the nut into one of the crankshaft keyways.

CRANKCASE COVER
Models 50-60-70

61. The crankcase cover can be removed for gasket renewal without removal of any other parts. If however, work is to be performed inside the crankcase, it is recommended that platform, hydraulic lines and "Powr-Trol" pump also be removed.

Fig. JD1058—The recommended crankshaft end play of 0.005-0.010 can be checked by using a dial indicator as shown.

OIL PRESSURE

Model 50

62. Recommended oil pressure is 10-15 psi when engine is running at high idle speed. To check and/or adjust the pressure, proceed as follows: Disconnect the oil pressure gage line (L—Fig. JD1060) from bushing (B) in main case near rear of governor housing and connect the master gage to the bushing. Start engine and observe oil pressure on master gage.

If pressure is not as specified, remove cap nut, loosen jam nut and turn adjusting screw **in** to increase pressure, **out** to decrease pressure as shown.

Models 60-70

62A. Recommended oil pressure is 10-15 psi when engine is running at high idle speed. To check and/or adjust the pressure, proceed as follows: Disconnect the oil pressure gage line (L—Fig. JD1061) from bushing (B) in rear of governor case and connect the master gage to the bushing. Start engine and observe oil pressure on master gage.

If pressure is not as specified, remove pipe plug from right side of main case and turn pressure regulating screw **in** to increase pressure and **out** to decrease oil pressure as shown.

FILTER HEAD & REGULATOR

Models 50-60-70

63. To remove the oil filter head, first drain the crankcase and remove the crankcase cover. Disconnect oil lines from filter head.

NOTE: If special tools are available, it is possible to disconnect these oil lines by working through the crankcase opening. Due to space limitations, it is often very time consuming, and if nipples connecting oil lines to filter head are not tight in filter head, it is oftentimes impossible for the average man to disconnect the oil lines by working through the crankcase opening; in which case, the following procedure is used.

63A. Disconnect connecting rods from crankshaft and remove carburetor, air inlet elbow and on models 50 and 60, remove tool box. Remove spark plug covers, disconnect spark plug wires and drain cooling system. Disconnect upper water pipe from cylinder block and remove lower water pipe. Unbolt cylinder block from main case and front end support. Slide cylinder head and cylinder block assembly forward as far as possible. Working through front opening in main case, disconnect the oil lines from filter head.

63B. After oil lines are disconnected, remove the filter element. Using 3/16 inch round, cold rolled steel rod, make up a tool similar to that shown in Fig. JD1062. Hook jaws of puller tool into holes of the filter outlet as shown in Fig. JD1063, then strike the bottom of the tool in a downward motion to remove the oil filter outlet. Working through bottom of filter body, remove the three cap screws retaining filter head to filter body.

Withdraw filter head and filter head cover through crankcase top opening. Make certain, however, that filter head assembly does not come apart and fall into crankcase.

Inspect filter head to make certain that mating and mounting surfaces of filter head are in good condition and not distorted. Check the condition of leaf springs and make sure that buttons in springs are tight. Oil pressure adjusting screws should turn freely. On models 60 and 70, the adjusting screw lock spring should have sufficient tension to hold adjusting screw in position.

When reassembling, use light gage wire to hold filter head and filter head cover together while installing the cap screws retaining the filter head to the filter body. Using a suitable piece of pipe, drive the filter outlet tube into position. When installing oil filter element, tighten the cover nut only enough to eliminate oil leakage.

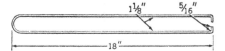

Fig. JD1062—Home made tool which can be used for removing the oil filter outlet as shown in Fig. JD1063. Tool can be made from 3/16 inch cold rolled steel rod.

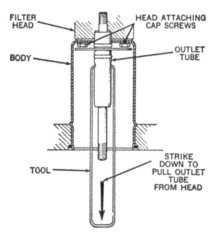

Fig. JD1063—Using the tool shown in Fig. JD1062 to remove models 50, 60 and 70 oil filter outlet. Use a hammer to strike down on tool.

Fig. JD1060—Checking and adjusting oil pressure on model 50 tractor. Turning screw in, increases the oil pressure.

Fig. JD1061—Checking and adjusting oil pressure on model 60. The method of adjusting the oil pressure on model 70 is similar. Pressure is increased by turning the screw in.

FILTER BODY

Models 50-60-70

64. Zero or low oil pressure can be caused by a distorted oil filter body (87—Fig. JD1066). Distortion is usually caused by over tightening the filter bottom retaining nut. The nut (84) should be tightened only to eliminate oil leakage at this point. If leakage occurs with normal tightening, renew gasket (81).

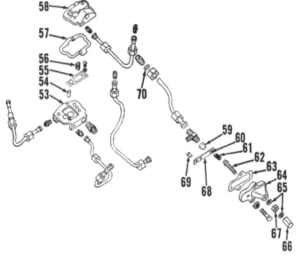

To renew the filter body, first remove the filter head as outlined in paragraph 63 and proceed as follows: Place a cylindrical wooden block in filter body and jack up the block, thereby forcing the filter body out of the crankcase recess.

To install a new oil filter body, coat the portion below the "bead" with white lead or equivalent sealing compound to prevent leaks and facilitate drawing body into recess of main case. Temporarily place filter head on filter

Fig. JD1064 — Exploded view of model 50 oil filter head and pipes. Oil pressure is adjusted externally with screw (62).

53. Filter head
54. Dowel pin
55. Leaf spring with button
56. Coil spring
57. Gasket
58. Filter head cover
59. Bushing
60. Relief spring pilot
61. Coil spring
62. Adjusting screw
63. Gasket
64. Housing
65. Copper washers
66. Cap nut
67. Jam nut
68. Leaf spring
69. Rivet
70. Washer

body to make certain that cap screw holes in body and head are aligned and oil lines will align with filter head connections; then lay filter head aside.

To draw filter body into crankcase recess, use a long bolt and two steel plates as shown in Fig. JD1067. Tighten the nut until bead of body seats against crankcase. Install the remaining parts by reversing the removal procedure.

OIL PUMP

Models 50-60-70

65. **BODY GEARS—RENEW.** To remove the body gears only, first drain crankcase and remove cover from pump body. Slide drive gear and shaft out of pump body as shown in Fig. JD1068. Remove crankcase cover and withdraw coupling (90—Fig. JD1069) from above. Pull drive gear and shaft from pump body.

Inspect the removed parts, using the values tabulated in paragraph 66A. Reassemble by reversing the disassembly procedure, making certain that coupling (90) engages the oil pump drive gear.

66. **R&R AND OVERHAUL.** To remove the complete oil pump, drain crankcase, remove crankcase cover and disconnect oil lines from oil pump body.

NOTE: If special tools are available, it is possible to disconnect these oil lines by working through the crankcase opening. Due to space limitations, it is often very time consuming, and

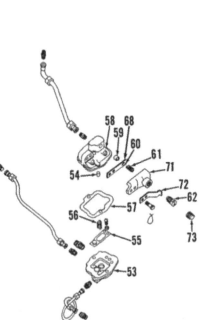

Fig. JD1065—Exploded view of models 60 and 70 oil filter head and pipes. Oil pressure is adjusted with screw (62).

53. Filter head	59. Bushing
54. Dowel pin	60. Relief spring pilot
55. Leaf spring with button	61. Coil spring
	71. Bracket
56. Coil spring	72. Adjusting screw spring
57. Gasket	73. Pipe plug
58. Filter head cover	

Fig. JD1066—Exploded view of models 50, 60 and 70 oil filter. When renewing element (75), tighten nut (84) only enough to eliminate oil leakage.

74. Filter outlet	81. Gasket
76. Snap ring	82. Cover
77. Spacer	83. Copper gasket
78. End plate and spacer	84. Nut
	85. Stud
79. Spring	86. Gasket
80. Washer	87. Oil filter body

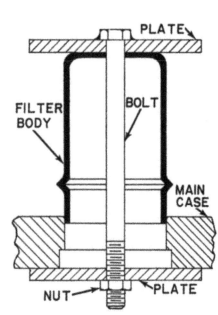

Fig. JD1067—Suggested home-made tool for installing models 50, 60 and 70 oil filter body. The long bolt can be welded to the upper plate.

if nipples connecting oil lines to pump are not tight in pump, it is oftentimes impossible for the average man to disconnect the oil lines by working through the crankcase opening; in which case, it will be necessary to move cylinder block forward and work through front opening in main case. Refer to paragraph 63A.

After the oil lines are disconnected, unbolt pump body from main case and withdraw pump assembly from below.

66A. Completely disassemble the pump and check the parts against the values which follow:

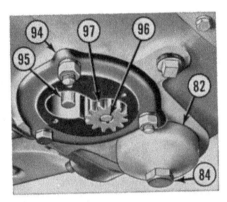

Fig. JD1068—Bottom view of models 50, 60 and 70 main case, showing oil pump and filter installation. The oil pump cover and idler gear have been removed.

Drive shaft bore in pump
  body 0.627-0.628
Drive shaft diameter.... 0.623-0.625
Diametral clearance between
  gear teeth and pump
  body 0.002-0.006
Gear bore in pump body. 2.086-2.088
Gear diameter 2.082-2.084
Idler gear shaft diam-
  eter 0.6285-0.6290
Idler gear shaft bore
  in pump body 0.627-0.628
Clearance between
  gears and cover 0.001-0.010
Depth of body gear bore. 1.240-1.245
Gear thickness 1.246-1.250
When reinstalling oil pump, make certain that coupling (90) engages the oil pump drive gear.

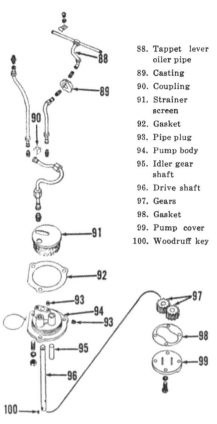

88. Tappet lever
    oiler pipe
89. Casting
90. Coupling
91. Strainer
    screen
92. Gasket
93. Pipe plug
94. Pump body
95. Idler gear
    shaft
96. Drive shaft
97. Gears
98. Gasket
99. Pump cover
100. Woodruff key

Fig. JD1069—Exploded view of models 50, 60 and 70 oil pump and associated pipes. Items (88) and (89) are located under tappet lever cover.

MANIFOLD
Models 50-60-70

67. Manifolds which are retained to cylinder head by cap screws can be removed without removing cylinder head from tractor. If manifold is retained by stud nuts, it is necessary to remove cylinder head before manifold can be removed.

CARBURETOR
(Not LP–Gas)

Models 50-60-70

68. Marvel-Schebler carburetors are used and their applications are as follows:

Model 50 Gasoline
  (Prior 5022299) .... DLTX 75

Other Model 50 Gas........ DLTX 86
Model 50 All-Fuel (Prior
  5015951) .... DLTX 73
Model 50 All-Fuel (After
  5015950) .... DLTX 83
Model 60 Gasoline ...... DLTX 81
Model 60 All Fuel (Prior
  6013900) .... DLTX 72
Model 60 All-Fuel (After
  6013899) .... DLTX 84
Model 70 Gasoline .... DLTX 82
Model 70 All-Fuel ...... DLTX 85

LP–GAS SYSTEM

Early models 60 and 70 were available with a single induction system, using Century equipment. Model 50 and late models 60 and 70 are available with a dual induction system, using a Deere carburetor and Century convertor and strainer.

Total fuel tank capacity is 39 gallons on models 60 and 70 and 28 gallons on model 50, but tank should NEVER be filled more than 85 per cent full of fuel (33 gallons on models 60 and 70 and 24 gallons on model 50). This allows room for expansion of the fuel due to a possible rise in temperature.

CAUTION: LP-Gas expands readily with any decided increase in temperature. If tractor must be taken into a warm shop to be worked on during extremely cold weather, make certain that fuel tank is as near empty as possible. LP-Gas tractors should never be stored or worked on in an unventilated space.

TROUBLE SHOOTING
Models 50-60-70

70. The following trouble shooting paragraphs list troubles that can be attributed directly to the fuel system; however, many of the troubles can be caused by derangement of other parts such as valves, battery, spark plugs, distributor, coil, resistor, etc.

The procedure for remedying many of the causes of trouble is evident. The following paragraphs will list the most likely causes of trouble, but only the remedies which are not evident.

Model	Float Setting	Repair Kit	Gasket Set	Inlet Needle and Seat	Nozzle	Economizer Plug	Power Jet or Needle
DLTX 72	⅜	286-877	16-662	X-2653	47-310	S 2151 G	......
DLTX 73	⅜	286-876	16-662	X-2653	47-314	S 2151 H	......
DLTX 75	*¾	286-996	16-671	233-543	47-362	......	49-191
DLTX 81	*¾	286-994	16-671	233-543	47-385	......	49-188
DLTX 82	*¾	286-1028	16-671	233-543	47-394	......	49-188
DLTX 83	*¾	286-1031	16-671	233-543	47-412	......	49-192
DLTX 84	*¾	286-1029	16-671	233-543	47-402	......	49-266
DLTX 85	*¾	286-1030	16-671	233-543	47-402	......	49-205
DLTX 86	*¾	286-1079	16-671	233-543	47-412	......	49-196

*Measured from nearest face of float to bottom of nozzle boss. On others, measured from bowl gasket seat in casting to bottom of float.

71. **HARD STARTING.** Hard starting could be caused by:

a. Improperly blended fuel.
b. Over-priming.
c. Incorrect starting procedure.
d. Primer not working properly. An audible click should be heard when the primer button is pushed. If no "click" is heard, check wiring and solenoid on back of convertor.
e. No fuel at carburetor. Open the liquid withdrawal valve, turn ignition switch on and press the primer button. If fuel is flowing to carburetor, an audible "hiss" will be heard. If no hiss is heard, refer to paragraph 81.
f. Automatic fuel shut-off or strainer not operating properly. A "click" should be heard when ignition switch is turned on. If no click is heard, check wiring and check solenoid on strainer.
g. Plugged vent on back of convertor. The vent is a ¼ inch tapped hole.
h. Defective low pressure diaphragm in convertor.
i. Binding carburetor metering valve or linkage.
j. Lean mixture caused by leaking gasket between two halves of carburetor on early models.

72. **ENGINE SHOWS NOTICEABLE LOSS OF POWER:**

a. Throttle not opened sufficiently due to maladjusted governor or carburetor linkage.
b. Plugged vent on back of convertor. The vent is a ¼ inch tapped hole.
c. Clogged fuel strainer (If strainer shows frost, it is probably clogged).
d. Plugged fuel lines or restrictions

in withdrawal valves (indicated by frost). With engine cold, both withdrawal valves closed and lines and filter empty of gas, remove plug at bottom of strainer, open the liquid withdrawal valve slightly and check for fuel flow.
e. Closed excess flow valves in vapor or liquid withdrawal valves (Indicated by frosted withdrawal valve). Close frosted valve to seat excess flow valve, then re-open slowly.
f. Lean mixture caused by restricted or altered fuel lines or hoses.
g. Sticking high pressure valve in convertor.
h. Restricted low pressure valve in convertor. Disconnect hose between carburetor and convertor, turn ignition switch on and press primer button to see if fuel is flowing.
i. Defective convertor diaphragms.
j. Faulty adjustment of carburetor drag link.
k. Binding carburetor metering valve.
l. Faulty adjustment of throttle rod.
m. Faulty carburetor load adjustment.
n. Faulty gasket between carburetor and manifold.
o. Leaking fuel hose between convertor and carburetor.
p. Air entering between carburetor throttle body and air horn.
q. Clogged air filter.
r. Lean mixture caused by leaking gasket between two halves of carburetor.

73. **POOR FUEL ECONOMY.** Could be caused by any of the conditions listed in paragraph 72, plus:

a. Improperly filled fuel tank.
b. Faulty fuel.
c. Faulty carburetor metering valve.

74. **ROUGH IDLING.** Could be caused by faulty ignition system plus:

a. Faulty adjustment of carburetor drag link.
b. Faulty adjustment of throttle rod.
c. Faulty carburetor to manifold gasket.
d. Leaking hose between carburetor and convertor.
e. Binding carburetor metering valve.

75. **POOR ACCELERATION.**

a. Faulty idle speed adjustment.
b. Faulty load adjustment.
c. Faulty low pressure diaphragm in convertor.
d. Restricted convertor to carburetor hose.
e. Leaking gasket between carburetor halves on early models.

76. **ENGINE STOPS WHEN THROTTLE IS BROUGHT TO SLOW IDLE POSITION.**

a. Faulty slow idle speed adjustment.
b. Faulty convertor to carburetor hose.
c. Faulty carburetor gaskets.
d. Faulty gasket between carburetor and air horn.
e. Leaking convertor back cover gasket (Indicated by fuel bubbles in radiator).

77. **OVERHEATING.** Could be caused by defective cooling system plus:

a. Lean mixture due to faulty ad-

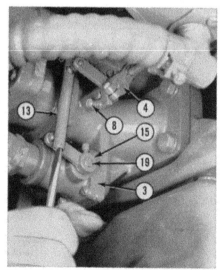

Fig. JD1076—Models 60 and 70 single induction LP-Gas carburetor installation. Synchronization of the metering valve and throttle disc is accomplished by adjusting the drag link as shown.

3. Metering valve   13. Drag link assembly
   housing   15. Metering valve
4. Throttle shaft   lever
8. Throttle stop screw 19. Metering valve

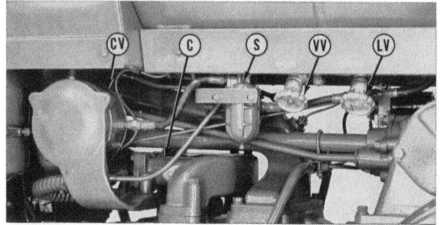

Fig. JD1075—Models 60 and 70 single induction LP-Gas tractor showing the installation of the fuel system components. Except for the carburetor, the dual induction systems are similar.

C. Carburetor   LV. Liquid withdrawal valve   S. Strainer and fuel shut-off
CV. Convertor   VV. Vapor withdrawal valve

justment of carburetor linkage and spray bar.

78. CONVERTOR FREEZES UP WHEN ENGINE IS COLD.

a. Running on liquid fuel before engine is warm.

b. Leaking convertor high pressure valve. With ignition switch turned on, this can be detected by odor of gas or hissing sound.

79. CONVERTOR FREEZES UP DURING NORMAL OPERATION.

Could be caused by a defective cooling system plus:

a. Water circulating backwards through convertor.

b. Restrictions in water piping or convertor.

c. Running on liquid fuel before engine is warmed up.

80. FROST ON WITHDRAWAL VALVE.

a. Closed excess flow valve. Close withdrawal valve to reset the excess flow valve; then, open withdrawal valve slowly.

b. Water in fuel tank will sometimes cause ice to form in liquid withdrawal valve. Empty all fuel from filler hose, pour one pint of alcohol in hose, attach hose to fuel tank and inject alcohol into fuel tank. Alcohol will act as an antifreeze and water will be dissipated through the engine.

81. LACK OF FUEL AT CARBURETOR.

Open the liquid withdrawal valve, turn ignition switch on and press the primer button. If fuel is flowing to carburetor, an audible "hiss" will be heard. If no hiss is heard, check the following possible causes.

a. Empty fuel tank or withdrawal valve closed.

b. Excess flow valve in withdrawal valve closed. Close the withdrawal valve to reset the excess flow valve; then, open the withdrawal valve slowly.

c. Restriction in withdrawal valve. See paragraph 80.

d. Restricted fuel strainer.

e. Faulty wiring to strainer shut off valve or faulty valve.

f. Faulty convertor high pressure valve.

g. Restricted fuel lines.

82. FUEL IN COOLING SYSTEM.

This can be checked by removing the radiator cap, turning on the ignition switch, pushing the primer button and watching for bubbles in the coolant. This trouble is usually caused by a ruptured convertor back cover gasket.

CARBURETOR
(Single Induction)
Models 60-70

83. ADJUSTMENT. Two speed adjustments and a load adjustment can be made on the carburetor and linkage. The proper adjustment of the slow idle speed synchronizes the carburetor metering valve and the throttle disc to provide the correct fuel-air ratio at all throttle settings.

84. SLOW IDLE ADJUSTMENT. Move the speed control hand lever all the way to the rear and turn the throttle stop screw (8—Figs. JD1076 and JD1079) to obtain a slow idle speed of 600 rpm when checked at belt pulley dust cover. Adjust the length of the speed control rod at the ball joint (Fig. JD1077) until the governor spring just touches the stop on the governor arm. Remove cotter pin from bottom of drag link and turn the adjusting screw as shown in Fig. JD1076 to obtain the fastest steady engine speed.

Reset the throttle stop screw (8), if necessary, to obtain an engine speed of 600 rpm and recheck the governor spring to be sure it just touches the governor arm.

85. FAST IDLE ADJUSTMENT. Move the speed control hand lever all the way forward and adjust cap screw (S—Fig. JD1077) to obtain a fast idle rpm of 1115.

86. LOAD ADJUSTMENT. If engine falters during acceleration, turn spray bar (2—Fig. JD1078) until arrow on spray bar points to second mark from the "R" (rich) mark on carburetor body. With engine idling, push the speed control hand lever forward quickly. If engine falters during this test, turn spray bar (2) slightly until arrow on spray bar points a little closer to "R" mark on carburetor body. Continue the adjusting procedure until engine accelerates smoothly.

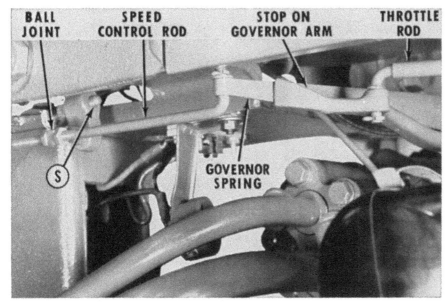

Fig. JD1077—Governor linkage adjustments on model 60 single induction LP-Gas tractor. The linkage on model 70 single induction LP-Gas tractor is similar. Fast idle speed is controlled by adjusting screw (S).

Fig. JD1078—Models 60 and 70 single induction LP-Gas carburetor load adjustment is accomplished by turning spray bar (2). The spray bar is a slight press fit in the carburetor body.

CARBURETOR
(Dual Induction)

Models 50-60-70

87. **ADJUSTMENT.** Two speed adjustments, two idle mixture adjusting needles and one load adjustment screw is provided on the carburetor and linkage.

87A. **SLOW IDLE SPEED ADJUSTMENT.** With the speed control hand lever in the closed (rearward) position and with the throttle closed, adjust the stop screw on carburetor throttle to provide a slow idle rpm of 300. Then adjust the length of the speed control rod (Fig. JD1079) until engine rpm is 600. Make certain, however, that the governor spring clears the stop screw during this adjustment.

Now, with engine running at 600 rpm, adjust the governor arm stop screw so that head of screw clears governor spring by $\frac{1}{8}$-inch as shown.

87B. **IDLE MIXTURE.** Both idle mixture needles (Fig. JD1079A) should be adjusted to provide steady operation throughout the idle speed range. Then, recheck the slow idle speed adjustment as in paragraph 87A.

87C. **FAST IDLE ADJUSTMENT.** With the speed control hand lever pushed all the way forward, adjust cap screw (S—Fig. JD1079) to obtain a fast idle (no load) speed of 1115 rpm for models 60 and 70, 1375 rpm for model 50.

87D. **LOAD ADJUSTMENT.** See Fig. JD1079B. Load adjustment screw on carburetor should be adjusted to provide the best operation under loaded conditions. Average adjustment is $1\frac{1}{4}$ turns open for models 60 and 70, $2\frac{1}{4}$ turns open for model 50.

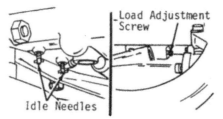

Fig. JD1079A—Models 50, 60 and 70 dual induction LP-Gas carburetor adjustment.

FUEL STRAINER AND SHUT-OFF
VALVE

Models 50-60-70

88. All of the fuel must pass through the strainer before reaching the convertor. The strainer contains a filter element which consists of a felt pad and a chamois disc backed by a brass screen on each side. The purpose of the filter is to remove all solids from the fuel before the fuel reaches the convertor valves. A solenoid operated, automatic fuel shut-off is located on top of the strainer. Whenever the ignition switch is turned on, the solenoid opens the valve with an audible "click."

If the strainer shows frost, it is probably clogged and needs cleaning.

89. **CLEANING.** To clean the strainer first make certain that both fuel tank withdrawal valves are closed, engine is cold and lines and filter are empty of gas. Note: Lines and filter will be empty if engine was properly stopped.

Remove plug (39—Fig. JD1080) from bottom of strainer and open the liquid withdrawal valve slightly; thus allowing pressure from the fuel tank to blow out any accumulation of dirt.

90. **R&R AND OVERHAUL.** To remove the strainer, close both fuel tank withdrawal valves, disconnect fuel lines and remove strainer.

Remove cover from strainer body and remove the filter pack by prying out the retainer ring with a screw driver as shown in Fig. JD1081. Filter pack can be cleaned in a suitable solvent. Reinstall filter pack with chamois disc toward top.

Disconnect wire and remove case from shut-off valve. Lift off coil (29—Fig. JD1082), remove plunger housing (30) and lift out plunger and spring. Inspect and renew any damaged parts.

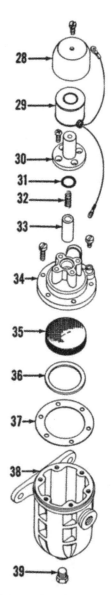

Fig. JD1080—Exploded view of models 50, 60 and 70 LP-Gas fuel strainer and automatic fuel shut-off.

28. Case	34. Strainer cover
29. Solenoid coil	35. Filter pack
30. Plunger housing	36. Retainer ring
31. "O" ring	37. Gasket
32. Spring	38. Strainer body
33. Plunger	39. Drain plug

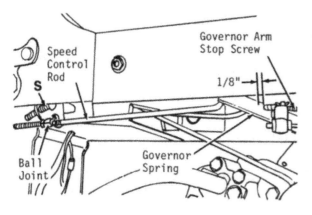

Fig. JD1079 — Governor linkage adjustment on models 50, 60 and 70 dual induction LP-Gas tractors. Fast idle speed is controlled by screw (S).

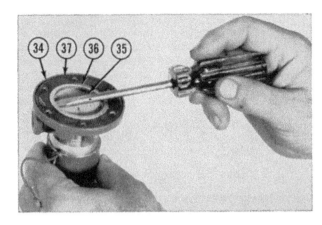

Fig. JD1081 — Removing filter pack retaining ring from LP-Gas fuel strainer cover.

34. Strainer cover
35. Filter pack
36. Retainer ring
37. Gasket

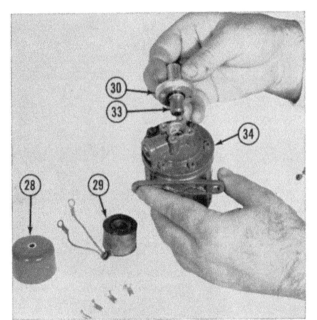

Fig. JD1082 — Removing the fuel shut-off plunger and plunger housing from LP-Gas fuel strainer cover.

28. Case
29. Solenoid coil
30. Plunger housing
33. Plunger
34. Strainer cover

Assemble the solenoid coil, plunger and spring on plunger housing.

To test solenoid, connect it to a battery to see if plunger compresses spring in plunger housing.

When reassembling, fasten one of the coil wires to screw which retains case (28). After unit is installed on tractor, test for leaks by using soapy water around all connections.

CONVERTOR
Models 50-60-70

91. **HOW IT OPERATES.** Liquid fuel (under relatively high pressure) passes through the solenoid operated fuel shut-off, the fuel strainer and into the convertor heat exchanger through the high pressure valve. Refer to Fig. JD1083. The liquid fuel is then vaporized by heat from the engine cooling system. Whenever the gas pressure reaches approximately 6 psi, the high pressure diaphragm spring is overcome and the high pressure valve is closed.

As the gas pressure exceeds 6 psi in the heat exchanger and as manifold vacuum reduces the pressure to below atmospheric in the low pressure chamber, the low pressure valve is opened and gas passes through the low pressure chamber and to the carburetor.

92. **OVERHAUL.** The high pressure valve can be removed and overhauled without removing convertor from tractor; any other work, however, cannot be accomplished until convertor has been removed.

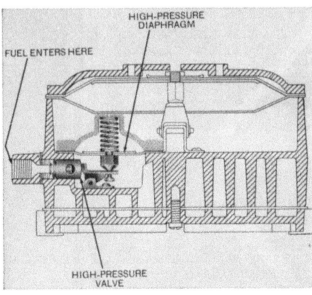

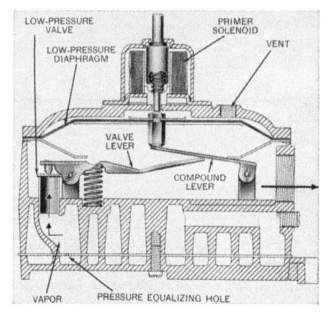

Fig. JD1083—Sectional views of LP-Gas convertor. Left view is sectioned through the high pressure system. Right view is sectioned through the low pressure system.

93. HIGH PRESSURE VALVE. To remove the high pressure valve for cleaning and/or parts renewal, as shown in Fig. JD1085, close both fuel tank withdrawal valves, disconnect the strainer-to-convertor fuel line and unscrew the high pressure valve jet from the convertor body.

Clean the valve assembly in a suitable solvent and renew any damaged parts. When reassembling, large end of spring (66—Fig. JD1084) goes toward aluminum seat (64). After valve is reinstalled, test for leaks with soapy water.

94. LOW PRESSURE DIAPHRAGM AND VALVE. With convertor removed from tractor, remove the convertor front cover (75—Fig. JD1084), low pressure diaphragm (72) and backing plate (69). Remove cotter pin from one end of primary lever pin (60) and remove the lever pin, primary lever (54), spring (62), low pressure valve seat (53) and low pressure valve (51) as shown in Fig. JD1086. Inspect all parts thoroughly and renew any which are excessively worn.

When renewing the neoprene valve seat (53—Fig. JD1084), hinge it loosely

to the primary lever so the seat can pivot on pin (52) and seat itself properly on the valve. Where reassembling, hook end of primary lever under the secondary lever and using a straight edge and rule, measure distance from end of secondary lever to edge of body casting as shown in Fig. JD1087. Bend the primary lever at (P), if necessary, until the measured distance is 1/8-5/32 inch as shown.

94A. Install diaphragm backing plate with "dish" down toward body and low pressure diaphragm with "dish" up toward cover. Bottom of front cov-

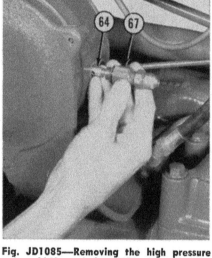

Fig. JD1085—Removing the high pressure valve (64) and jet (67) from LP-Gas convertor.

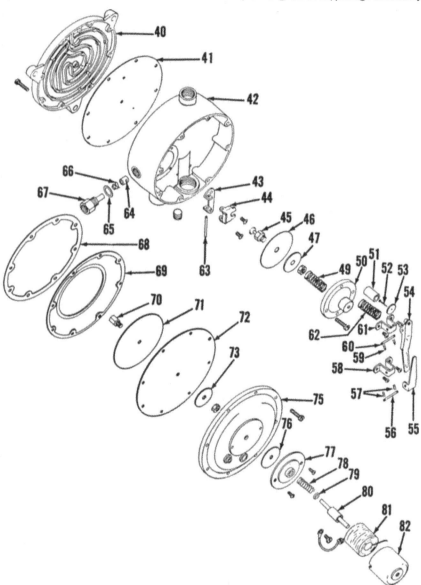

Fig. JD1084—Exploded view of LP-Gas convertor. The solenoid operated primer is used for starting only.

40. Back cover	51. Low pressure valve	60. Pivot pin
41. Gasket	52. Pin	61. Primary lever
42. Convertor body	53. Low pressure	bracket
43. High pressure valve	valve seat	62. Spring
operating bracket	54. Low pressure valve	63. Pivot pin
44. Lever	primary lever	64. High pressure valve
45. Diaphragm link	55. Secondary lever	65. Gasket
46. High pressure	56. Pivot pin	66. Spring
diaphragm	57. Hair pin	67. High pressure jet
47. Diaphragm plate	58. Low pressure valve	68. Gasket
49. Spring	lever bracket	69. Backing plate
50. Cover	59. Hair pin	70. Diaphragm button

71. Diaphragm plate
72. Low pressure	diaphragm
73. Diaphragm plate
75. Front cover
76. Gasket
77. Primer base
78. Spring
79. "O" ring
80. Plunger
81. Solenoid coil
82. Case

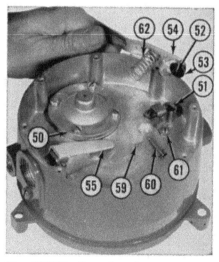

Fig. JD1086—Removing the low pressure valve primary lever from LP-Gas convertor.

51. Low pressure valve		55. Secondary lever
52. Pin		59. Hair pin
53. Low pressure		60. Pivot pin
valve seat		61. Primary lever
54. Low pressure valve		bracket
primary lever		62. Spring

er goes toward convertor attaching lug which is partially ground off. NOTE: Before completely tightening the cover cap screws, push down on the solenoid plunger to make certain that low pressure diaphragm has sufficient sag to operate properly. Also, make certain that there are no wrinkles along edge of diaphragm.

95. HIGH PRESSURE DIAPHRAGM. With convertor removed from tractor, remove the convertor front cover (75 —Fig. JD1084), low pressure diaphragm (72) and backing plate (69). Remove the high pressure diaphragm cover and lift off the diaphragm with spring and link as shown in Fig. JD1088.

Thoroughly inspect all parts and renew any which are damaged. Bleed hole in top of diaphragm cover should be open and clean.

When reassembling, hook button on diaphragm link (45) under operating lever (44) and install spring and valve cover. Make certain that there are no wrinkles along edge of diaphragm before tightening the cover screws. Install low pressure diaphragm and front cover as in paragraph 94A.

96. HEAT EXCHANGER. If convertor shows frost when engine is warm, and if the pipes carrying water to and from the convertor are not plugged, remove convertor from tractor and take off the back cover. Thoroughly clean the water chamber

and renew the back cover gasket. Bleed hole in gasket goes toward top of convertor and milled-off attaching lug on cover goes toward bottom.

97. CONVERTOR SOLENOID PRIMER. If primer is inoperative (an audible "click" should be heard each time primer button is pressed), remove the convertor from tractor and take off the complete solenoid primer. Check to make certain that plunger (80—Fig. JD1084) is not binding due to foreign material and check solenoid with a 12-volt battery. If primer still does not operate, disassemble and renew any questionable parts.

Fig. JD1088 — Removing high pressure valve diaphragm from LP-Gas convertor. Bleed hole in cover should be open and clean.

44. Lever
45. Diaphragm link
46. High pressure diaphragm
49. Spring
50. Cover

When reassembling, place coil ground wire between coil and top cover (50) to assure a good ground connection.

GOVERNOR

Models 50, 60 and 70 are equipped with a centrifugal flyweight type governor which is driven by the engine camshaft gear. An idler gear, which is mounted in the rear portion of the governor case drives the live "Powr-Trol" pump. The fan drive pinion is mounted on the governor shaft, and is in constant mesh with the fan drive bevel gear. Refer to Fig. JD1090.

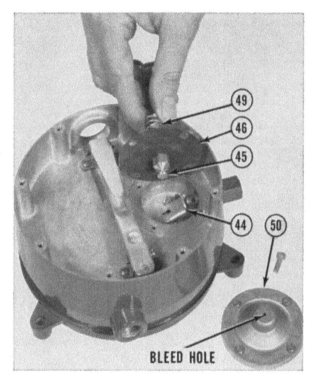

BLEED HOLE

Fig. JD1090 — Sectional view of models 60 and 70 governor. Model 50 is similar except the flyweights are mounted on the governor drive gear.

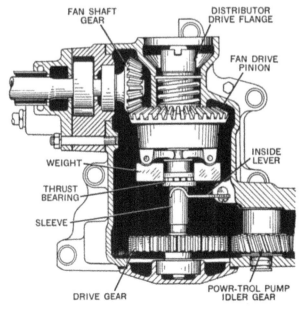

FAN SHAFT GEAR

DISTRIBUTOR DRIVE FLANGE

FAN DRIVE PINION

WEIGHT

INSIDE LEVER

THRUST BEARING

SLEEVE

DRIVE GEAR

POWR-TROL PUMP IDLER GEAR

$\frac{1}{8} = \frac{5}{32}''$

Fig. JD1087—Models 50, 60 and 70 LP-Gas convertor. Bend primary lever (54) at point (P) so that distance from end of secondary lever (55) to edge of convertor body is 1/8-5/32 inch as shown.

SPEED AND LINKAGE ADJUSTMENT

Models 50-60-70

100. Before attempting to adjust the engine speed, free-up and align all linkage to remove any binding tendency and adjust or renew any parts causing lost motion. Cap screw (8—Fig. JD1091) should be adjusted so that a pull of ten pounds at end of speed control lever is required to move lever through full range of travel when rod (6) is disconnected from speed control arm (11). Adjust the length of the throttle rod (4) so that rod is ½ hole short when speed control lever and throttle butterfly are in wide open position. This check should be made at governor end of rod.

On early models, adjust the throttle stop screw on carburetor to limit the slow idle engine speed to 600 rpm. Move the throttle lever to wide open position and turn screw (12—Figs. JD-1091 or 1091A) to limit the fast idle, no load engine speed to 1115 rpm for models 60 and 70, 1375 rpm for model 50. Full load engine speed should be 975 rpm for models 60 and 70, 1250 rpm for model 50. Engine speed can be checked at belt pulley dust cover.

Late production gasoline and distillate engines are fitted with the same linkage arrangement as the late production of LP-Gas tractors. Refer to Fig. JD1079. On this late style linkage, follow the adjustment procedure in paragraphs 87A and 87C.

OVERHAUL

Models 50-60-70

101. *Normal overhaul of the governor consists of removing and overhauling the shaft assembly only; and can be accomplished without removing governor housing from tractor. If, however, the fan drive bevel gear is damaged it will be necessary to remove the complete governor assembly as well as the fan shaft assembly in order to renew the matched set of bevel gears. Refer to paragraph 102 or 103 for renewal of bevel gears.*

101A. **SHAFT AND WEIGHTS.** To overhaul the governor shaft and weights, first loosen set screw (1—Fig. JD1091) and bump governor arm up and off the governor lever shaft. Remove bearing housing (17—Figs. JD1092 or 1093) and save shims (18) for reinstallation. Turn flywheel until governor weights are on top and bottom (vertical position) and withdraw governor shaft assembly as shown in Fig. JD1094.

Inspect all parts for damage or excessive wear. Inspect the bearing on

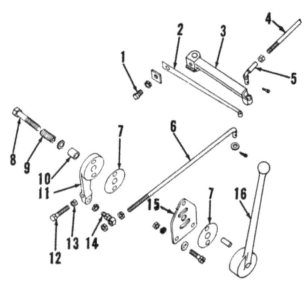

Fig. JD1091 — Exploded view of models 50, 60 and 70 governor controls. The fast idle speed is controlled by limiting screw (12).

1. Set screw
2. Governor spring
3. Governor arm
4. Throttle rod
5. Rod end
6. Speed control rod
7. Friction washer
8. Adjusting screw
9. Spring
10. Spacer
11. Speed control arm
12. Speed adjusting screw
13. Jam nut
14. Ball joint
15. Speed control plate
16. Speed control lever

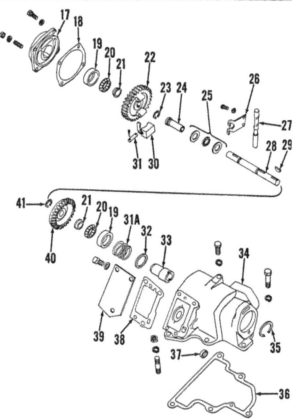

Fig. JD1091A—Model 70 speed control linkage, showing the location of speed adjusting screw (12). The construction is similar on models 50 and 60.

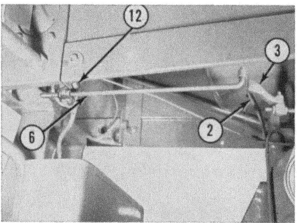

Fig. JD1092 — Exploded view of model 50 governor assembly. Backlash between the fan drive bevel gears is controlled by shims (18). Items (37 and 39) are not used if tractor is equipped with "Powr-Trol."

17. Left bearing housing
18. Shim gaskets
19. Bearing cup
20. Balls and retainer
21. Bearing cone
22. Drive gear
23. Snap ring
24. Sleeve
25. Thrust bearing
26. Governor lever
27. Lever shaft
28. Governor shaft
29. Woodruff key
30. Flyweight
31. Pin
31A. Bearing spring
32. Washer
33. Distributor drive flange
34. Housing
35. Snap ring
36. Gasket
37. Plug
38. Gasket
39. Cover
40. Fan drive bevel pinion
41. Snap ring

right end of shaft for evidence of turning in governor case. If bearing has been turning in governor case, check the case by installing a new bearing cup. If a new cup fits loosely, it will be necessary to renew the governor case or reclaim same as in paragraph 104. Distributor drive flange (33—

Figs. JD1092 or JD1093) can be removed from governor shaft by using a suitable puller. If the fan drive bevel pinion is damaged, press governor shaft out of pinion and renew the bevel gears which are available in a matched set only.

NOTE: If necessary to renew the bevel gears, it will also be necessary to remove the fanshaft assembly and adjust the mesh and backlash of the bevel gears as outlined in paragraph 108.

The spur drive gear (22) can be pulled or pressed from governor shaft if renewal is required. Thrust bearing (25) should be in good condition and sleeve (24) should slide freely on the governor shaft.

The governor lever shaft (27) should have a clearance of 0.002-0.004 in governor case. Bearing spring (31A) should require 99-121 pounds to compress it to a height of 1 1/16 inches. NOTE: Spring test does not apply to a reworked spring which is mentioned in paragraph 104.

When reassembling, use Figs. JD1092 or 1093 as a guide. Press the fan drive bevel pinion on governor shaft (if removed) until shoulder on pinion contacts snap ring on shaft. Press new drive gear on shaft until shoulder on gear contacts snap ring on shaft. On models 60 and 70, long hub of drive gear goes toward left end of shaft.

101B. Install governor shaft assembly by reversing the removal procedure and mesh the "V" marked tooth on governor gear with the similarly marked tooth space on camshaft gear as shown in Fig. JD1095. When gears are properly timed, the slot in the ignition distributor drive flange and the slot in flywheel hub will be parallel to floor. If fan drive bevel pinion was not renewed, install same number and thickness of shims (18—Figs. JD1092 or 1093) as were originally removed so as to retain the desired bevel gear backlash of 0.004-0.007 for models 50 and 60, 0.004-0.006 for model 70. NOTE: Removal of ignition distributor will facilitate installation of governor shaft. Refer to paragraph 113 for ignition timing.

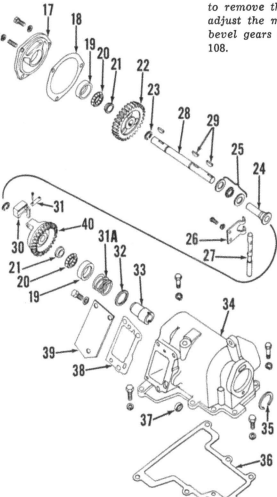

Fig. JD1093 — Explanded view of model 70 governor. Model 60 governor is similarly constructed. Items (37 and 39) are not used if tractor is equipped with "Powr-Trol." Refer to legend for Fig. JD1092.

Fig. JD1094—Removing model 50 governor shaft. Notice that weights are in vertical position. Models 60 and 70 are similar except flyweights are mounted on the fan drive bevel pinion.

Fig. JD1095—Models 50, 60 and 70 governor shaft is properly installed when the "V" marked drive gear tooth is meshed with the similarly marked tooth space on the camshaft gear.

REMOVE AND REINSTALL

As previously outlined in paragraph 101, the governor housing need not be removed to facilitate a normal overhaul of the governor shaft and weights. The subsequent paragraphs, however, will outline the procedure for removing the governor housing assembly to perform any of the following jobs.

A. Removal of engine camshaft and associated parts.

B. Removal of fan shaft assembly.

C. Renewal of fan drive bevel gears.

D. Reclaiming of governor case.

Model 50

102. To remove governor housing assembly, first unscrew starter pedal button and remove flywheel cover and flywheel. Drain the hydraulic "Powr-Trol" system and remove platform, hydraulic lines and hydraulic pump. Disconnect wires from ignition distributor and remove the distributor. Disconnect speed control rod from governor spring and throttle rod from governor arm. Unbolt the fanshaft rear bearing housing from governor housing and governor housing from main case, noting the position of the dowel type cap screws. Disconnect oil pressure line from bushing in main case, located at right rear of governor housing. Raise governor until governor gear and camshaft gear are out of mesh, move governor rearward until the fan drive bevel gears are out of mesh and withdraw governor assembly from tractor.

NOTE: Be careful not to lose the fan drive bevel gear mesh adjusting shims which are located between the fanshaft rear bearing housing and the governor case. If the bevel gears are not to be renewed, the same shims should be used during reassembly.

When reinstalling governor, refer to paragraph 101B. Adjust crankshaft end play as outlined in paragraph 59 and ignition timing as outlined in paragraph 113.

Models 60-70

103. To remove the governor housing assembly, remove grille and loosen the two cap screws retaining front of hood to the radiator top tank. Remove the four cap screws retaining instrument panel to support, disconnect oil pressure line from gage, pull instrument panel rearward and remove the two cap screws retaining the steering shaft and hood support to the gear shift quadrant. On late models so equipped, disconnect oil lines from the automatic fuel shut off valve located at top of fuel filter. On all models, disconnect oil line from connector located at right rear of governor case. Disconnect speed control rod from governor spring, throttle rod from governor arm, choke rod and fuel line from carburetor, coil wire from coil and fuel tank support from governor housing. On model 70 it is also necessary to unbolt the fuel tank support from the upper water pipe rear casting. On all models, raise hood and fuel tank assembly and block-up between the steering shaft and hood support and the gear shift quadrant.

Drain the hydraulic "Powr-Trol" system and remove the platform, hydraulic lines and hydraulic pump. Disconnect wires from ignition distributor and remove the distributor. Unscrew packing gland which retains the ventilator pipe to the vent pump cover and unbolt the fanshaft rear bearing housing from the governor housing. Unbolt front of fanshaft from its support and remove the "U-Bolt" which retains the spark plug wire conduit to the fanshaft tube.

Unbolt governor housing from main case, raise governor until governor gear and camshaft gear are out of mesh, move governor rearward until the fan drive bevel gears are out of mesh and withdraw governor assembly from tractor. Refer to Fig. JD1096.

NOTE: Be careful not to lose the fan drive bevel gear mesh adjusting shims which are located between the fanshaft rear bearing housing and the governor case. If the bevel gears are not to be renewed, the same shims should be used during reassembly.

When reinstalling governor, refer to paragraph 101B. Adjust crankshaft end play as outlined in paragraph 59 and ignition timing as outlined in paragraph 113.

RECLAIMING GOVERNOR CASE
Models 50-60-70

104. If the governor shaft right hand bearing cup has been turning in governor housing, and if a new bearing cup fits loosely in the housing bore, it will be necessary to renew housing or reclaim same. To reclaim housing, remove and disassemble governor and proceed as follows:

Using a Woodring and Wise reclaiming fixture as shown in Fig. JD1097, ream the governor housing bearing bore to an inside diameter of 2.247 and to a depth at which the back side of the reamer is flush with inner edge of bearing boss. Obtain a John Deere F152R bushing and cut the bushing to a width of ⅝ inch. Press bushing in governor housing and ream the bushing to an inside diameter of 2.080. Reassemble governor by reversing the disassembly procedure.

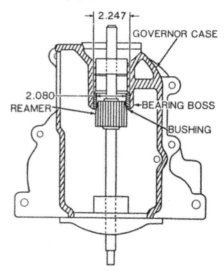

Fig. JD1097—Using a Woodring and Wise reclaiming fixture on models 50, 60 and 70 governor case. See text. The fixture is available from Woodring and Wise Machine Co., Waterloo, Iowa.

Fig. JD1096—Governor housing partially removed from model 60 tractor. Although the flywheel is off in this illustration, it is not necessary to remove flywheel to R & R governor.

An alternate reclaiming procedure is as follows:

Grind 19/32 inch from inner end of distributor drive flange as shown in

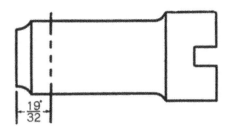

Fig. JD1098. NOTE: Bearing cone No. JD7656R is 19/32 inch wide. Shorten the governor shaft bearing spring (31A—Figs. JD1092 or 1093) by 19/32 inch. This can be accomplished by grinding.

Reassemble governor as follows: Press old bearing cone on governor shaft with small diameter of cone toward bevel pinion as shown in Fig. JD1099. Press new bearing cone on governor shaft with larger diameter of cone toward bevel pinion. Install the remaining parts of the bearing, shortened spring, washer and shortened distributor drive flange as shown.

Fig. JD1098—The distributor drive flange must be shortened 19/32 inch when reclaiming the governor case by the alternate method given in the text.

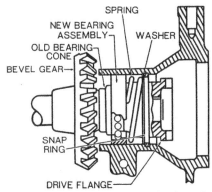

Fig. JD1099—Sectional view showing the proper location of parts in governor housing when the housing has been reclaimed by the alternate method given in the text.

COOLING SYSTEM

RADIATOR

Models 50-60

105. **REMOVE AND REINSTALL.** To remove the radiator, it is first necessary to remove the hood and fuel tank assembly as follows: Drain cooling system and remove grille, steering wheel and Woodruff key. Remove the steering wormshaft front bearing housing and turn the wormshaft forward and out of housing; OR, unbolt worm housing from pedestal and remove wormshaft and housing assembly from tractor. Disconnect choke rod at carburetor and remove fuel line. On late models, disconnect oil lines from the automatic fuel shut-off valve which is located on top of fuel strainer. Remove muffler. Disconnect battery cable, remove the four cap screws retaining instrument panel to support, disconnect oil line from oil pressure gage and pull instrument panel rearward. Disconnect coil and regulator wires at instrument panel. Unscrew pull knob from choke rod and pull choke rod forward until free from rear support. Unclip heat indicator bulb wire from hood. Unbolt hood and lift hood and fuel tank assembly from tractor.

Remove air intake and exhaust pipes and loosen the air cleaner hose clamps. The air cleaner can be removed for convenience. On early models so equipped, remove the radiator shutter. Unbolt baffle plates from the radiator lower tank. Loosen fan belt and unbolt water pump from lower tank. Unbolt radiator from front end support and using a hoist as shown in Fig. JD1100, lift radiator assembly from tractor.

Model 70

106. **REMOVE AND REINSTALL.** To remove the radiator, drain cooling system and remove grille, steering wheel and Woodruff key. Remove the

Fig. JD1100—Using a chain hoist to lift models 50, 60 and 70 radiator assembly from tractor.

Fig. JD1101 — Cut-away view of crankcase ventilator pump used on models 50, 60 and 70. The pump rotor is mounted on the rear of the fan shaft and the pump housing is bolted to the governor case.

steering wormshaft front bearing housing and turn the wormshaft forward and out of housing; OR, unbolt worm housing from pedestal and remove wormshaft and housing assembly from tractor. Remove muffler, unbolt hood and lift hood from tractor.

Remove air intake and exhaust pipes and loosen the air cleaner hose clamps. The air cleaner can be removed for convenience. Loosen fan belt and unbolt water pump from lower tank. Unbolt radiator from front end support and using a hoist as shown in Fig. JD1100, lift radiator assembly from tractor.

FAN SHAFT AND VENTILATOR PUMP

Models 50-60-70

The fan shaft is driven by a bevel pinion which is mounted on the governor shaft (Refer to Fig. JD1090). The engine crankcase is ventilated by a rotor-type pump which is mounted on the rear of the fanshaft and bolted to the governor case (Refer to Fig. JD1101).

107. R&R AND OVERHAUL. To remove the fan shaft assembly on models without power steering, first remove governor as outlined in paragraph 102 or 103 and proceed as follows: Remove the lower half of the carburetor air intake casting. Loosen the air intake hose clamps and remove the upper half of the air intake casting and the crankcase vent pipe. Remove exhaust pipe, loosen generator and roll generator out toward right side of tractor. On model 50, unbolt fan shaft from its front support and on model 70, unbolt the fan shaft front support from tractor frame, being careful not to lose spacer washers which are located between fan support and frame.

On all models, withdraw fan shaft assembly, front end first, from right side of tractor.

After fan shaft assembly is removed from tractor, inspect the fan drive bevel gear (70—Fig. JD1102). If the

Fig. JD1103—When disassembling models 50, 60 and 70 fan shaft, compress the assembly enough to remove the half moon locks (62) and keeper (61). After releasing the pressure, the remaining parts can be removed.

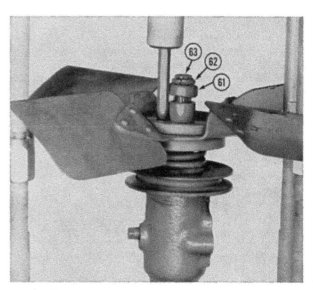

bevel gear must be renewed, refer to paragraph 108.

To disassemble the removed fan shaft, place the assembly in a press and remove the half-moon locks (62) and keeper (61) as shown in Fig. JD1103. Remove the assembly from press, disassemble and check the remaining parts against the values which follow:

Fan blade pitch (Inches)
    Models 50 & 60....2 11/32-2 13/32
    Model 70 ....2 17/32-2 19/32

Thickness of friction washers
    Models 50-60-70 ....1/16 inch

Friction spring strength
    Model 50 ..112-138 lbs. @ 1½ inches
    Model 60 ..171-209 lbs. @ 1½ inches
    Model 70 ..216-264 lbs. @ 1½ inches

Bearing take-up spring strength
    Model 50 .70-86 lbs. @ 1 7/16 inches
    Models 60 and 70
         171-209 lbs. @ 1 7/16 inches

Ventilator pump specifications are as follows:

Roller diameter 0.4982-0.4988
Roller length 0.965-0.970
Rotor thickness 0.964-0.965
Rotor diameter 2.425-2.430
Rotor groove diameter...0.500-0.505
Pump body thickness .0.9715-0.9735

Use Fig. JD1102 as a general guide during reassembly and be sure to renew all "O" ring packings and felt washers.

Note: On models with power steering, the procedure for removing the fan shaft is evident after removing the power steering pump as in paragraph 20.

Fig. JD1102—Exploded view of models 50, 60 and 70 fan shaft and crankcase ventilator pump as used on models with manual steering. Shim gaskets (71) control mesh position of the fan drive bevel gears. On models with power steering, refer also to Fig. JD1014.

45. Washer	54. Felt retainer	62. Half-moon locks
46. Packing	55. "O" ring	63. Fan shaft
47. Retainer	56. Pulley	64. Key
48. Spring	57. Spring	65. Front bearing
49. Bearing cup	58. Friction disc	housing and tube
50. Balls & retainer	59. Friction washer	66. Pipe plug
51. Bearing cone	60. Fan drive cup	67. "O" ring
52. Washer	61. Fan keeper	68. Vent pump roller
53. Felt washer		

69. Snap ring	
70. Bevel gear	
71. Shim gasket	
72. Fan shaft rear	
bearing housing	
(pump housing)	
73. Pump rotor	
74. Gasket	
75. Vent pump cover	

FAN DRIVE BEVEL GEARS

Models 50-60-70

The fan drive bevel gears (40 & 70—Fig. JD1104) are available in a matched pair only. Therefore, if either gear is damaged, it will be necessary to renew both gears and adjust them for the proper mesh and backlash.

108. To renew the bevel gears, remove governor as outlined in paragraph 102 or 103 and fanshaft as outlined in paragraph 107. Remove the governor shaft assembly from governor housing and using a suitable puller, remove the distributor drive flange and bevel pinion from governor shaft. Press new bevel pinion on shaft until pinion seats against snap ring. Reassemble governor shaft and reinstall in governor housing.

Using a suitable press, press the fan drive bevel gear (70—Fig. JD1102) further on fan shaft until snap ring (69) can be removed. Remove snap ring and bevel gear and install the new bevel gear.

Before installing fan shaft or governor assembly on tractor, temporarily bolt fanshaft to governor case, observe mesh position of bevel gears and add or remove shim gaskets (71—Fig. JD1104) until heels of gears are in register. Mount a dial indicator in a manner similar to that shown in Fig. JD1105 and check the bevel gear backlash. Add or remove shim gaskets (18—Fig. JD1104) until backlash is 0.004-0.007 for models 50 and 60, 0.004-0.006 for model 70.

After proper backlash is obtained, unbolt fan shaft from governor housing and save the mesh adjusting shim gaskets for installation when fan shaft and governor are reinstalled on tractor.

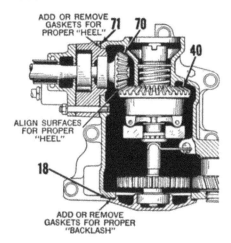

Fig. JD1104—Sectional view of models 60 and 70 governor, showing points for adjusting the fan drive bevel gears.

18. Shims
40. Bevel pinion
70. Bevel gear
71. Shims

THERMOSTAT

Models 50-60-70

109. The desired operating temperature is automatically maintained on early model 50 and 60 tractors by a radiator shutter which is actuated by a thermostat located in the radiator lower tank as shown in Fig. JD1106. The thermostat is accessible for removal after removing the grille and disconnecting the shutter linkage.

When reinstalling the thermostat, check the shutter operating linkage for proper adjustment as follows: With engine cold and shutter completely closed, link (86—Fig. JD1107) should slide freely into hole of bell crank (84). If adjustment is required, disconnect bell crank from adjustable connector (79) and turn the connector in or out as required.

110. On models not covered by paragraph 109, the thermostat is located in the upper water pipe rear casting (cylinder block water outlet casting). The procedure for removing the thermostat is evident.

Fig. JD1105—Checking backlash of the fan drive bevel gears on model 50 tractor. The same procedure can be used on models 60 and 70.

Fig. JD1106—Cut-away view of model 60 cooling system showing the thermostat (78) installation on models equipped with a radiator shutter (88). The water pump is shown installed at (P).

WATER PUMP

Coolant leakage at drain hole in pump housing usually indicates a leaking seal or cracked carbon thrust washer.

Models 50-60-70

111. **R&R AND OVERHAUL.** To remove the water pump on models 50 and 60, drain cooling system and disconnect fan belt and lower hose from pump. Remove the two cap screws retaining the fan shaft to the fan shaft front support and unbolt the support from tractor frame. CAUTION: Do not lose shims located between the fan support and tractor frame. Move the fan support rearward, unbolt water pump from radiator and withdraw pump from tractor.

To remove the water pump on model 70, drain cooling system, disconnect fan belt from pump and remove the lower water pipe. Unbolt and withdraw water pump from tractor.

On all models, use a suitable puller and remove the drive pulley. Remove snap ring (90—Fig. JD1108) and press shaft and bearing assembly forward out of impeller. The bellows seal and carbon washer can be renewed at this time. The shaft and bearing (91) are available as an assembled unit only. If the small brass slinger on pump shaft

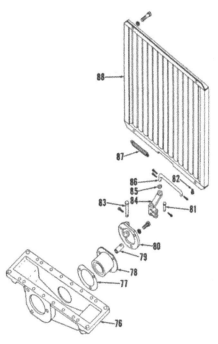

Fig. JD1107—Exploded view of thermostat, shutter and associated linkage as used on some 50 and 60 models.

76. Radiator lower tank
77. Gasket
78. Thermostat
79. Adjustable connector
80. Thermostat cover
82. Rivet
83. Pin
84. Bell crank
85. Washer
86. Control link
87. Spring
88. Shutter assembly

is renewed, press slinger on shaft until edge of slinger is 1 39/64 inches from end of shaft as shown in Fig. JD1109.

When reassembling, install the shaft and bearing unit in pump housing and install snap ring (90 — Fig. JD1108). On model 60, install the

right hand pump attaching cap screw (one nearest pump outlet). On all models, support pump shaft from underneath side and press pulley on shaft until pulley is flush with end of shaft. Coat sealing surfaces with water-resistant grease and assemble the bellows seal (94) and carbon

thrust washer (93) in impeller, making certain that lugs on thrust washer fit into slots of impeller. Place impeller on flat surface and press body and shaft assembly onto impeller until highest vane on impeller is flush with mounting surface of pump body.

NOTE: If any vane protrudes beyond the mounting face of pump housing, the protruding vane will strike the radiator lower tank when pump is installed.

Install pump on tractor by reversing the removal procedure.

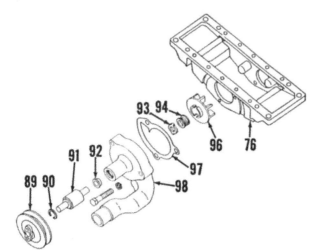

Fig. JD1108 — Exploded view of model 50 water pump. Models 60 and 70 are similarly constructed except model 70 pump is retained by four cap screws.

89. Pulley
90. Snap ring
91. Shaft and bearing assembly
92. Slinger
93. Carbon washer
94. Bellows seal
96. Impeller
97. Gasket
98. Pump housing

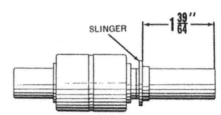

Fig. JD1109—The brass slinger should be pressed on models 50, 60 and 70 water pump shaft until slinger is 1 39/64 inches from end of shaft as shown.

IGNITION AND ELECTRICAL SYSTEM

DISTRIBUTOR
Models 50-60-70

112. APPLICATIONS. Model 50 tractors prior to Ser. No. 5016500 were factory equipped with a Wico model XB4023 battery ignition unit which can be replaced with the later production Delco-Remy model 1111558 distributor. The LP-Gas models 50, 60 and 70 are equipped with a Delco-Remy model 1111418. All other models are equipped with a Delco-Remy model 1111558.

The 1111418 distributor, used on the LP-Gas models, rotates at ½ crankshaft speed. All other models rotate at crankshaft speed.

Distributor Test Specifications

DR 1111418
  Contact Gap 0.022
  Cam Angle 54-62
  Rotation, Drive End C
  Advance Data .... 0-2.5@200
                           1.5-4@250
                            3-6@300
                            7-9@400

DR 1111558
  Contact Gap 0.022
  Cam Angle 86-96
  Rotation, Drive End CC
  Advance Data .... 1-5@400
                            10-14@700
                            18-22@900
                            23-26@1000

Wico XB4023
  Contact Gap 0.015
  Rotation, Drive End CC
  Starting Retard .... 25°

113. INSTALLATION AND TIMING. The distributor can be timed in either the fully advanced or full retard position. Normally, the distributor is installed and timed in the full retard position; then the running timing is checked and the necessary slight adjustments made at rated engine speed. Breaker contact gap is 0.015 for Wico units, 0.022 for Delco-Remy.

To install and time the distributor, loosen cap screw and slip cover away from the timing port in flywheel housing. Crank engine until number one (left) cylinder is coming up on compression stroke and the static ignition timing mark on flywheel rim is in register with index in timing port. The static timing mark is as follows:

Models 50 & 60 Gasoline .. 5° ATDC
Model 70 Gasoline ...... 10° ATDC
Models 50, 60 & 70 All
  Fuel .... TDC
Models 60 & 70 LP-Gas ... 10° ATDC
Model 50 LP-Gas .... 5° ATDC

On the LP-Gas models 60 and 70, hold distributor in the position shown in Fig. JD1111, with coil wire terminal toward tractor and bottom cover clamp straight down. Turn rotor to the 7:30 o'clock position and mount distributor making certain that the driving gears mesh properly. Drain hole in distributor cap goes toward bottom and number one terminal is away from tractor. Rotate top of distributor away from tractor and turn on the

ignition switch. Slowly tap top of distributor toward tractor until a spark occurs for number one cylinder. Tighten the distributor mounting cap screws at this point.

On all other models, the slot in the distributor drive flange should be approximately parallel to ground as shown in Fig. JD1112 when the static ignition timing mark on flywheel rim is in register with index in the inspection port. If slot is not parallel to ground, the governor gear and camshaft gear are not meshed properly (Refer to paragraph 101B). Turn the distributor shaft until rotor arm is in the number one firing position (toward top terminal in distributor cap). Install the distributor and tighten the

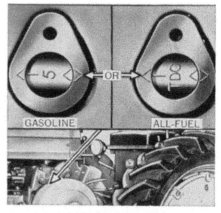

Fig. JD1110—Static ignition timing marks on models 50 and 60. Static timing marks on other models are given in text.

mounting cap screws finger tight. Rotate top of distributor toward front of tractor and turn on the ignition switch. Slowly tap top of distributor toward rear until a spark occurs for number one cylinder and tighten the distributor mounting cap screws.

Check the ignition running timing with a timing light as shown in Fig. JD1113. Desired running timing is as follows:

Models 50 & 60 Gasoline...20° BTC
Model 70 Gasoline........15° BTC
Models 50, 60 & 70 All Fuel.25° BTC
Models 60 & 70 LP-Gas.... 5° BTC
Model 50 LP-Gas ....10° BTC

GENERATOR AND REGULATOR
Models 50-60-70

114. Delco-Remy generators and regulators are used on all models. Refer to the actual unit for model number.

Test specifications are as follows:

Generator 1100955
Brush spring tension ....16 oz.
Field draw, volts 12.0
   amps 2.0-2.14

Hot output, amps 9.0-11.0
   volts 13.8-14.2
   rpm 2400

Generator 1100309
Brush spring tension........28 oz.
Field draw, volts 12.0
   amps 1.58-1.67
Cold output, amps 20.0
   volts 14.0
   rpm 2300

Regulators 1118306, 1118783 & 1118792
Cutout relay, air gap........0.020
   point gap 0.020
   closing voltage range..11.8-14.0
   adjust voltage to 12.8
Regulating unit, air gap 0.075
   voltage range 13.6-14.5
   adjust voltage to 14.0

STARTING MOTOR
Models 50-60-70

115. Delco-Remy starting motors are used on all models. Refer to the actual unit for model number. Test specifications are as follows:

Fig. JD1111 — Installing the ignition distributor on LP-Gas models 60 and 70. The unit rotates at ½ engine speed and is driven by a pair of spiral gears.

Models 1108144 & 1108155
Volts 12
Brush spring tension ..24-28 ounces
No load test, volts 11.3
   amps 70
   rpm 6000
Lock test, volts 6.7
   amps 530
   torque, Ft.-Lbs. 16

Models 1108950, 1108989, 1108981 & 1108990
Volts 12
Brush spring tension ..36-40 ounces
No load test, volts 11.3
   amps 65
   rpm 5500
Lock test, volts 4.0
   amps 675
   torque, Ft-Lbs. 30

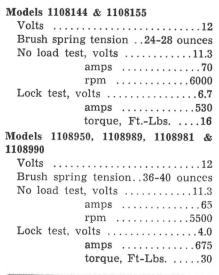

Fig. JD1112—Model 70 governor with the distributor removed. Notice that drive slot (S) in coupling is parallel to ground.

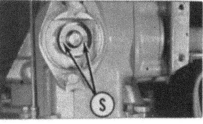

Fig. JD1113—Fully advanced ignition timing marks on model 50. Advanced timing marks on other models are given in text.

CLUTCH, BELT PULLEY AND PULLEY BRAKE

ADJUSTMENT
Models 50-60-70

116. To adjust the clutch, remove the belt pulley dust cover and the cotter pin from each of the three clutch operating bolts. Place the clutch operating lever in the engaged position (lever fully forward) and tighten each adjusting nut (21—Fig. JD1120) a little at a time and to the same tension. Check tightness of clutch after each adjustment by disengaging and reingaging clutch. When the adjustment is correct, a distinct snap will occur when the clutch is engaged and 40-60 lbs. pressure on models 50 and 60 or 60-80 lbs. on model 70

quired at the end of the operating lever to lock the clutch in the engaged position with engine running at idle speed.

To adjust the pulley brake on model 50 prior to Ser. No. 5029200, model 60 prior to Ser. No. 6054500, and model 70 prior to Ser. No. 7031300, proceed as follows: Note: These early models

Fig. JD1120—Models 50, 60 and 70 clutch is adjusted by turning the three castellated adjusting nuts (21) until 40-60 lbs. pressure on models 50 and 60 or 60-80 lbs. pressure on model 70 at end of clutch lever is required to engage clutch when engine is idling.

have the headless operating pin (41— Figs. JD1125 and 1126). Engage clutch and hold pulley brake tightly against

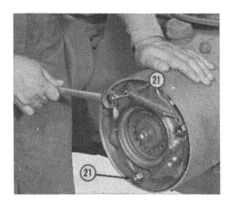

belt pulley (engine stopped). Loosen lock nut (39—Figs. JD1121, 1123A, 1125 or 1126) and turn adjusting screw (38) until there is a clearance of approximately $\frac{1}{16}$-$\frac{1}{8}$ inch between end of adjusting screw and the operating pin (41—Figs. JD1125 or 1126) when the operating pin is pushed against the clutch fork.

To adjust the pulley brake on model 50 after Ser. No. 5029199, model 60 after Ser. No. 6054499 and model 70 after Ser. No. 7031299, proceed as follows: Note: These late models have the headed operating pin shown in Fig.

Fig. JD1121—Models 50 and 60 pulley brake installation. The brake is adjusted by loosening jam nut (39) and turning adjusting screw (38).

JD1127. With the pulley brake adjusting screw loosened, engine running and clutch disengaged, turn the adjusting screw inward until the pulley will just stop turning and tighten the adjusting screw jam nut.

116A. If transmission is hard to shift or if gears clash, the position of the headed pulley brake operating pin must be adjusted as follows: Remove the clutch fork bearing and vary the number of shims, as required, under the pin head to just prevent clutch operating sleeve from striking the pulley gear when the clutch lever is pulled back. This will be a trial and error job from bench to tractor.

After final assembly and adjustment of clutch and with pulley brake adjusting screw fully loosened, the pulley should have $\frac{1}{16}$-$\frac{1}{8}$ inch end play on crankshaft. If end play is excessive, add more shims under the operating pin head and recheck. When end play is as specified, readjust pulley brake.

RENEW CLUTCH FACINGS
Models 50-60-70

117. To remove the clutch discs and facings, remove the pulley dust cover and adjusting disc (30—Figs. JD1123 or 1124). Withdraw the lined and unlined discs. Remove cap screw (25) retaining the clutch drive disc (22)

Fig. JD1123A—Model 70 belt pulley and pulley brake installation. The pulley brake adjustment is similar to models 50 and 60.

to crankshaft and using jack screws or a suitable puller, remove the clutch drive disc and inner facing.

Worn, badly glazed or oil soaked facings should be renewed. A facing that is in usable condition, is quite rigid. Any facing that bends easily should be renewed. Renew release springs (32—Fig. JD1123 or 1124) if they are rusted, distorted or do not meet the following specifications:

Pounds Test @ Height

Model 50 .20-30 lbs. @ 1 5/16 inches
Model 60 ...45-55 lbs. @ 1⅝ inches
Model 70 ...45-55 lbs. @ 1⅝ inches

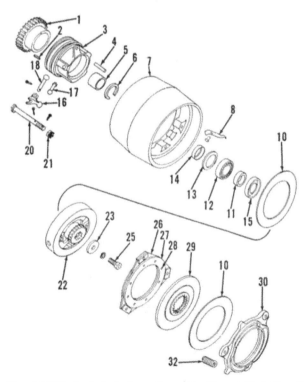

Fig. JD1123—Exploded view of models 50 and 60 belt pulley and clutch assembly. The clutch is adjusted with nuts (21).

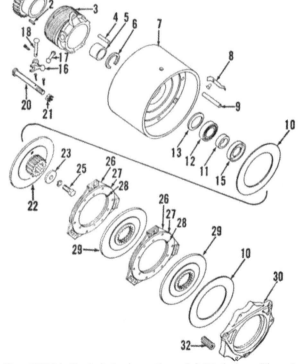

Fig. JD1124—Exploded view of model 70 belt pulley, clutch and associated parts. A counter weight on the engine crankshaft compensates for the relatively light drive disc (22).

1. Drive gear	7. Pulley	12. Bearing	20. Operating bolt	27. Facing
2. Key	8. Spring	13. Bearing washer	21. Adjusting nut	28. Rivet
3. Operating sleeve	9. Drive pin (70)	14. Oil retainer (50)	22. Clutch drive disc	29. Sliding drive disc
4. Drive pin	10. Clutch facing	15. Bearing retainer	23. Washer	30. Adjusting disc
5. Bushing	11. Bearing inner race	16. Clutch dog	25. Cap screw	32. Spring
6. Snap ring	(60 & 70)	17. Dog toggle	26. Facing disc	

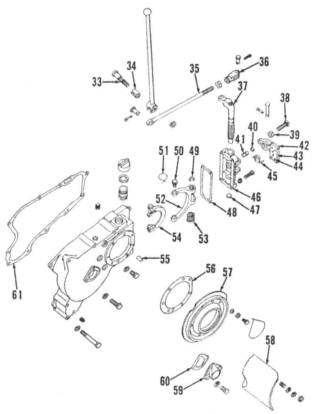

Fig. JD1125—Exploded view of early model 50 reduction gear cover, belt pulley brake and associated parts. Early model 60 construction is similar. On later models, pin (41) is headed and requires a shim adjustment.

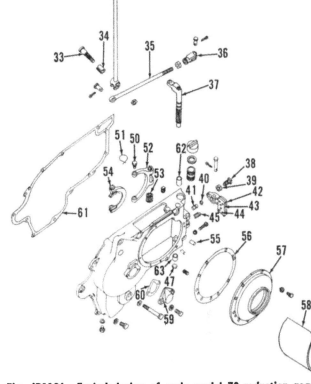

Fig. JD1126—Exploded view of early model 70 reduction gear cover, belt pulley brake and associated parts. On later models, pin (41) is headed and requires a shim adjustment.

33. Pivot bolt	40. "O" ring	45. Spring	52. Clutch fork	58. Pulley guard
34. Bushing	41. Pulley brake	46. Clutch fork bearing	53. Spring	59. Cover
35. Operating rod	operating pin	(50 & 60)	54. Clutch collar	60. Gasket
36. Yoke	42. Pulley brake	47. Expansion plug	55. Dowel pin	61. Gasket
37. Clutch fork shaft	43. Brake lining	48. Gasket (50 & 60)	56. Gasket	62. Bushing (70)
38. Adjusting screw	44. Rivet	49. Snap ring (50 & 60)	57. Dust shield	63. Lower bushing (70)

Use Fig. JD1123 or 1124 as a guide during reassembly and on models 50 and 60, install the clutch drive disc so that "V" mark on drive disc is in register with "V" mark on crankshaft as shown in Fig. JD1128. Adjust the clutch as outlined in paragraph 116.

R&R BELT PULLEY
Models 50-60

118. To remove the belt pulley, first remove the clutch facings as outlined in paragraph 117 and disconnect the clutch operating rod from the clutch fork shaft. Unbolt and remove the pulley brake, clutch fork shaft, bearing and clutch fork as an assembly from tractor. Withdraw pulley from crankshaft.

Model 70

119. To remove the belt pulley, first remove the clutch facings as outlined in paragraph 117 and disconnect the clutch operating rod from the clutch fork shaft. Remove pivot pin and withdraw pulley brake (42 — Fig. JD1126). Unbolt cover (57) from re-

duction gear case and withdraw pulley assembly from tractor.

OVERHAUL PULLEY
Models 50-60-70

120. To disassemble the removed pulley, use a punch and hammer as shown in Fig. JD1129 and remove the

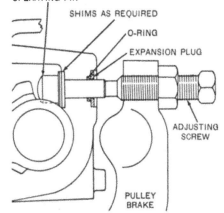

Fig. JD1127—Top sectional view of late production clutch fork bearing, showing the installation of the shim adjusted pulley brake operating pin.

bearing retainer and bearing (12—Fig. JD1123 or 1124). Remove the operating bolts (20), pins (18), dogs (16) and toggles (17). Using a suitable puller, remove drive gear (1) and slide operating sleeve from pulley.

Fig. JD1128—When installing the clutch drive disc on models 50 and 60, make certain that "V" mark on the clutch drive disc is in register with the "V" mark on end of crankshaft.

Check pulley bushing (5) and the engine crankshaft against the values which follow:

Crankshaft diameter at pulley bushing
Model 50 .... 1.994-1.996
Model 60 .... 2.245-2.247
Model 70 .... 2.432-2.434

Suggested clearance between pulley bushing and crankshaft
Models 50-60 .... 0.003-0.007
Model 70 .... 0.0035-0.0065

If clearance between pulley bushing (5) and crankshaft is excessive, renew the bushing. Using a piloted drift, press bushing into pulley until bushing seats against snap ring (6). If bushing is carefully installed, no final sizing will be required. Diameter of the installed bushing should be not less than 1.999 for model 50, 2.250 for model 60 and 2.4375 for model 70.

When reassembling, install gear (1) with long hub toward pulley.

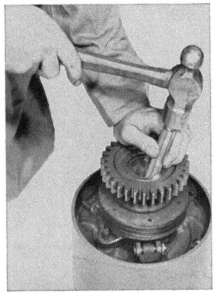

Fig. JD1129—Using a punch and hammer to remove the pulley bearing on models 50, 60 and 70.

CLUTCH CONTROLS

Models 50-60

121. **R&R AND OVERHAUL.** The clutch fork (52—Fig. JD1125) and clutch collar (54) can be renewed after disconnecting the clutch operating rod, unbolting the clutch fork shaft bearing from the reduction gear cover and withdrawing bearing and fork assembly from tractor. The procedure for subsequent disassembly is evident. Check the component parts against the following:

Thickness of collar (54) . 0.307-0.312
Diameter of pivot holes
   in collar (54) 0.499-0.501
Diameter of upper hole in fork
   shaft bearing (46) ... 1.0635-1.0665
Diameter of lower hole in
   bearing (46) 0.811-0.814
Diameter of fork shaft (37)
   at lower end 0.807-0.809
Diameter of fork shaft (37)
   at upper end 1.060-1.062

Fig. JD1130 — Model 70 clutch fork (52) and collar (54) installation as viewed after pulley has been removed. Pin (41) operates the pulley brake.

when reassembling, refer to paragraph 116A.

Model 70

122. The clutch fork (52—Figs. JD1126 and 1130) and the clutch collar (54) can be renewed after removing the belt pulley as outlined in paragraph 119. The procedure for subsequent disassembly is evident. Check the component parts against the values which follow:

Thickness of collar
   (54) 0.3695-0.3745
Diameter of pivot holes in
   collar (54) 0.499-0.501
Diameter of fork shaft (37)
   at lower end 0.743-0.745
Diameter of fork shaft (37)
   at upper end 1.060-1.062
Min. clearance between fork
   shaft and upper bushings.... 0.002
Min. clearance between fork
   shaft and lower bushing..... 0.003

When reassembling, refer to paragraph 116A.

TRANSMISSION

The transmissions used in models 50, 60 and 70 provide six forward speeds and one reverse.

The tractors may be equipped with either a continuous (live) or a non-continuous power take-off. Power for the continuous power take-off originates with a spur gear which is keyed to the engine crankshaft and is not dependent on any of the transmission shafts or gears. The non-continuous (transmission driven) power take-off is driven by the transmission countershaft and is actuated by shifting a sliding gear into mesh with the countershaft idler gear.

The sliding gear shaft drive shaft and integral gear are driven by the first reduction gear which meshes with a spur gear on the belt pulley.

The differential spur gear meshes with the differential drive pinion which is located on the transmission countershaft.

OVERHAUL

Model 50

123. **FIRST REDUCTION GEAR COVER.** To remove the first reduction gear cover, disconnect the clutch operating rod and drain oil from cover. Remove belt pulley as outlined in paragraph 118. Move the right rear

wheel out and remove the right brake assembly as outlined in paragraph 152. Remove the transmission drive shaft right bearing cover (87—Fig. JD1131), extract cotter pin and remove nut (49) from end of shaft. Remove cap screws retaining the reduction gear cover to main case and bump end of drive shaft with a soft hammer as shown in Fig. JD1132 to loosen the reduction gear cover. When cover is free from dowels, withdraw cover, front end first, as shown in Fig. JD1133.

When reassembling, soak new reduction gear cover gasket until gasket is pliable, shellac gasket to main case and install reduction gear cover by reversing the removal procedure.

124. TRANSMISSION TOP COVER. To remove the transmission top cover and shifter quadrant assembly, proceed as follows: Disconnect battery cable and oil gage line. Remove the four cap screws retaining instrument panel to the steering shaft rear support, pull instrument panel rearward and disconnect wires which go through the gear shifter quadrant. Remove both cap screws retaining the steering shaft rear support to the gear shifter quadrant, raise steering shaft support and rear end of hood approximately 1½ inches and block up between hood and governor housing. Remove the cap screws retaining the shifter quadrant and transmission cover assembly to main case and with gear shift lever on right side of quadrant, withdraw top cover and shifter quadrant assembly from tractor.

125. SHIFTER SHAFTS AND SHIFTERS. To remove the shifter shafts and shifters, first remove the engine flywheel as outlined in paragraph 59 and the transmission top cover as outlined in paragraph 124. The fourth and sixth speed shifter shaft pawl and spring should be removed at this time. They are located in a vertical drilled hole in main case to the left of the top opening and are retained by the transmission top cover. Using a pry bar through transmission top opening, move each shifter along its shaft until the detent pawls rise and hold the pawls in the raised position by inserting a cotter pin or wire in the exposed hole in each pawl as shown in Fig. JD1135. Unwire and remove set screw (26—Fig. JD1136) which positions the fourth and sixth speed shifter on its shaft.

Fig. JD1131 — Rear sectional view of model 50 main case, showing the installation of the transmission shafts and gears. The unit shown is not equipped with continuous power take-off.

A. Powershaft sliding gear
B. Powershaft bevel pinion
40A. Spacer
45. First reduction gear
49. Nut
52. Transmission drive shaft
54. Sliding gear shaft drive gear
55. First and third sliding pinion
56. Second and fifth sliding pinion
57. Sliding gear shaft
63. Fourth and sixth drive gear
65. Fourth and sixth sliding pinion
71. Second and fifth speed gear
72. Differential drive pinion
74. First and third speed gear
76. Bearing housing
84. Countershaft idler gear
86. Countershaft
87. Cover

Fig. JD1134—Exploded view of model 50 shifter quadrant and transmission cover assembly. Spring (8) should test 50 lbs. when compressed to a height of 15/16-1 1/16 inch.

1. Quadrant	8. Spring
2. Transmission cover	9. Ball socket cover
3. Gasket	10. Fulcrum seal
4. Shift ball	11. Gasket
5. Lock washer	12. Ball socket
6. Snap ring	13. Fulcrum ball
7. Washer	14. Gear shift lever

Fig. JD1132—The reduction gear cover can be loosened from main case by tapping end of drive shaft with a soft hammer.

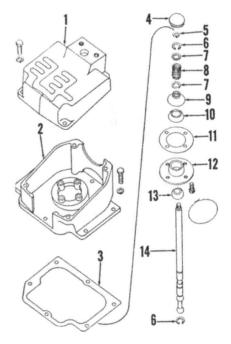

Fig. JD1133—Removing model 50 reduction gear cover. Front of cover must be withdrawn first to permit rear of cover to clear the first reduction gear and powershaft idler.

Fig. JD1135—Top view of model 50 transmission with top cover removed. Notice how pawls (35) are locked in the raised position by using pieces of wire.

15. Set screw
16. Fourth & sixth speed shifter fork
17. Fourth & sixth pawl
18. Locking cap screw (three used)
19. Safety wire
20. Pawl spring
21. Adjusting screw (three used)
22. Second and fifth shifter shaft
23. Fourth and sixth shifter shaft
24. First, third and reverse shifter shaft
25. Underdrive shifter shaft
26. Drilled set screw
28. Underdrive shifter
29. Underdrive shifter yoke
30. Lock plate
31. Jam nut
32. Stop screw
33. First, third and reverse shifter
34. Pawl spring (three used)
35. Pawl (three used)
36. Second and fifth speed shifter
37. Fourth and sixth speed shifter

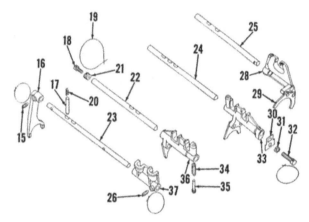

Fig. JD1136 — Exploded view of model 50 shifter shafts and shifters. Stop screw (32) and lock plate (30) are accessible after removing the reduction gear cover.

Remove the fourth and sixth speed gear cover which is located under flywheel on left side of main case. Remove cap screw (18—Fig. JD1136 or 1137) and adjusting screw (21) from left end of each shifter shaft. Pull each shifter shaft toward left to disengage it from locking plate (30 — Fig. JD1136) which holds the right end in place, rotate shafts sufficiently to move detents out of alignment with pawls, withdraw shifter shafts from left and shifters from above. Remove the rear shifter shaft and shifter first, then work forward, removing the remaining shafts and shifters.

New shifter yokes can be riveted to shifters if old yokes are worn or bent. Renew any shifter shaft that is worn around the detent. Renew any pawl that is worn out-of-round at ball end. The first and third as well as the second and fifth speed shifter springs should test 42-51 lbs. when compressed to 1¾ inches. The fourth and sixth speed shifter spring should have a free length of 1-1¼ inches.

When reinstalling the shifter shafts and shifters, refer to Fig. JD1136 and reverse the removal procedure, making certain that flat on right end of shifter shafts engage locking plate (30). After the shifter shafts and shifters are installed, place the fourth and sixth speed shifter in neutral position, making certain that pawl engages detent in shaft. Turn the adjusting screw (21), located at left end of

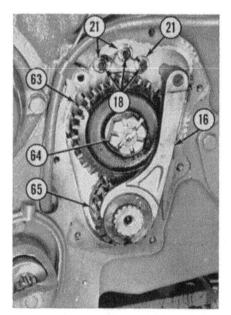

Fig. JD1137—Left side of model 50 main case with fourth and sixth gear cover removed.

16. Fourth & sixth speed shifter fork
18. Locking cap screw
21. Adjusting screw

Fig. JD1138 — Rear sectional view of model 50 transmission as used on tractors with transmission driven power take-off. An idler gear replaces spacer (40A) when tractor is equipped with engine driven (continuous) power take-off.

A. Powershaft sliding gear
B. Powershaft bevel pinion
40A. Spacer
45. First reduction gear
49. Nut
52. Transmission drive shaft
54. Sliding gear shaft drive gear
55. First and third sliding pinion
56. Second and fifth sliding pinion
57. Sliding gear shaft
63. Fourth and sixth drive gear
65. Fourth and sixth sliding pinion
71. Second and fifth speed gear
72. Differential drive pinion
74. First and third speed gear
76. Bearing housing
84. Countershaft idler gear
86. Countershaft
87. Cover

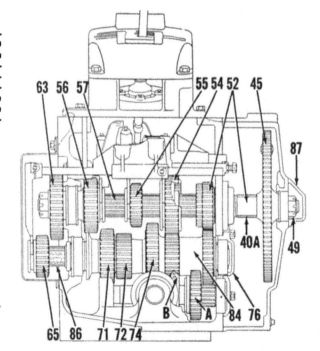

shafts (22, 24 and 25), in or out, until left hand shifter gates are aligned. After the adjustment is complete, install and tighten the locking cap screws (18) securely.

Move shifter (33) to first or third speed position; at which time, there should be a gap of approximately 5/64 inch between end of stop screw (32) and right end of shifter (33), and the first and third speed drive and sliding gears should be meshed properly.

NOTE: If stop screw (32) is screwed in far enough to prevent first and third speed gears from going into full mesh or if gap between stop screw and shifter is excessive enough to permit more than ⅛ inch overshift, it will be necessary to remove clutch, belt pulley and reduction gear cover as in paragraph 123 to permit readjustment of the stop screw.

126. SLIDING GEAR SHAFT. To remove the sliding gear shaft (57—Fig. JD1138) and gears, remove the transmission top cover as in paragraph

Fig. JD1140—Right side of model 50 main case with the reduction gear cover, first reduction gear and power shaft idler gear removed.

124 and the shifter shafts and shifters as outlined in paragraph 125. Remove cotter pin and nut (64—Fig. JD1137) from left end of sliding gear shaft and using a suitable puller, remove the fourth and sixth speed drive gear (63). Remove the sheet metal oil retainer (62—Fig. JD1139), extract snap ring (59) and pull the sliding gear shaft toward left until the left bearing (60) emerges from the main case.

Bump pilot bearing (53) from right end of shaft with sliding drive gear (54), withdraw sliding gear shaft from left side of case and remove gears from above.

Inspect all parts and renew any which are questionable. When reassembling, pack pilot bearing (53) with wheel bearing grease and install in drive shaft (52) with shielded side of bearing toward left side of tractor. Install sliding gear shaft and gears by reversing the removal procedure and install snap ring (59) with gap in snap ring spanning the oil passage in main case. Install the sheet metal oil retainer (62) with flat spot adjacent to oil passage in main case. After installing the fourth and sixth speed drive gear, tighten nut (64) securely.

127. TRANSMISSION DRIVE SHAFT. To remove the drive shaft (52—Figs. JD1138 and 1139), remove the clutch, belt pulley and first reduction gear cover as outlined in paragraph 123 and the sliding gear shaft as outlined in the preceding paragraph 126. Withdraw the first reduction gear (45) and spacer (40A) if tractor is equipped with a transmission driven power shaft, or idler gear (40) if engine driven power shaft is used. Remove bearing cover (38—Fig. JD1140) and withdraw drive shaft from tractor.

When installing the shaft, refer to Fig. JD1139 as a reference and reverse the removal procedure.

128. COUNTERSHAFT. To remove the countershaft (86—Figs. JD1138 and 1141), remove the sliding gear shaft as in paragraph 126 and the transmission drive shaft as outlined in the preceding paragraph. Remove the countershaft right bearing housing (76—Fig. JD1140) and unstake and re-

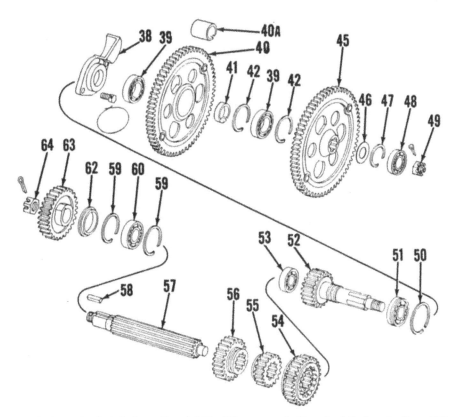

Fig. JD1139—Exploded view of model 50 sliding gear shaft and related parts. Gear (40) is the engine driven powershaft idler. Spacer (40A) is used when tractor is equipped with a transmission driven power shaft.

38. Bearing cover	48. Bearing	56. Second and fifth sliding pinion
39. Bearing	49. Nut	57. Sliding gear shaft
40. Powershaft idler gear	50. Snap ring	58. Key
40A. Spacer	51. Bearing	59. Snap ring
41. Spacer	52. Transmission drive shaft	60. Bearing
42. Snap ring	53. Pilot bearing	62. Oil retainer
45. First reduction gear	54. Sliding gear shaft drive gear	63. Fourth and sixth drive gear
46. Washer	55. First and third sliding pinion	64. Nut
47. Snap ring		

move nut (66—Fig. JD1141) from left end of shaft. Using a soft hammer, bump countershaft out right side of case and remove gears from above.

The countershaft right bearing cup can be pulled from bearing housing (76) if renewal is required and the left bearing cup can be driven from the main case bore.

When reassembling, use Fig. JD1141 as a guide, install the same number of shims (77) as were originally removed and tighten nut (66) securely. Stake the nut into one of the shaft splines. Mount a dial indicator as shown in Fig. JD1142 and check the shaft end play which should be 0.001-0.004. If end play is not as specified, remove

the countershaft right bearing housing (76—Fig. JD1141) and add or remove the required amount of shims (77).

Model 60

130. **FIRST REDUCTION GEAR COVER.** To remove the first reduction gear cover, disconnect the clutch operating rod and drain oil from cover. Remove belt pulley as outlined in paragraph 118. Move the right rear wheel out and using a double nut arrangement, remove the upper right hand implement mounting stud from the right front face of the rear axle housing. Unbolt and withdraw the right brake assembly. Remove the transmission drive shaft right bear-

ing cover (82—Fig. JD1143), extract cotter pin and remove nut (44) from end of shaft. Remove cap screws retaining the reduction gear cover to main case and bump end of drive shaft with a soft hammer as shown in Fig. JD1144 to loosen the reduction gear cover. Pry cover from its locating dowels and withdraw the first reduction gear cover from tractor. Be careful not to lose the camshaft end play removing spring which is retained by the reduction gear cover.

When reassembling, soak new reduction gear cover gasket until gasket is pliable, shellac gasket to main case and install reduction gear cover by reversing the removal procedure.

131. **TRANSMISSION TOP COVER.** To remove the transmission top cover and shifter quadrant assembly, remove grille and loosen the two cap screws retaining front of hood to the radiator top tank. Remove the four cap screws retaining instrument panel to support, disconnect oil pressure line from gage, pull instrument panel rearward and remove the two cap screws retaining the steering shaft and hood support to the gear shift quadrant. Disconnect battery cable and disconnect wires from instrument panel which go through the gear shifter quadrant. On late models so equipped, disconnect oil lines from the automatic fuel shut off located at top of fuel filter. On all models, disconnect oil line from connector located at right rear of governor case. Disconnect speed control rod from governor

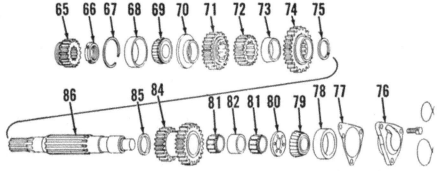

Fig. JD1141—Exploded view of model 50 transmission countershaft and gears.

65. Fourth and sixth speed sliding pinion	71. Second and fifth speed gear
66. Lock nut	72. Differential drive pinion
67. Snap ring	73. Spacer
68. Bearing cup	74. First and third speed gear
69. Bearing cone	75. Snap ring
70. Spacer	76. Bearing housing
	77. Shims & gaskets
	78. Bearing cup

79. Bearing cone
80. Thrust washer
81. Bearing
82. Spacer
84. Idler gear
85. Collar
86. Countershaft

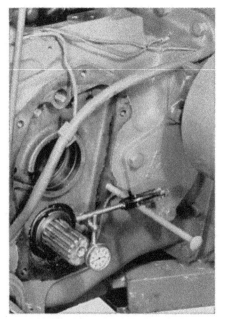

Fig. JD1142—Using a dial indicator to check end play of model 50 transmission countershaft. The recommended end play of 0.001-0.004 is controlled by shims under the right bearing housing.

Fig. JD1143 — Rear sectional view of model 60 transmission with non-continuous power take-off. An idler gear replaces spacer (37A) on models equipped with engine driven power shaft.

A. Powershaft sliding gear
B. Powershaft drive pinion
37A. Spacer
40. First reduction gear
44. Nut
45. Fourth and sixth sliding gear
53. Transmission drive shaft
56. Sliding gear shaft
57. First and third sliding pinion
58. Second and fifth sliding pinion
59. Sliding gear shaft drive gear
62. Fourth and sixth drive gear
66. First and third gear
68. Differential drive pinion
69. Second and fifth gear
72. Countershaft
74. Countershaft idler gear

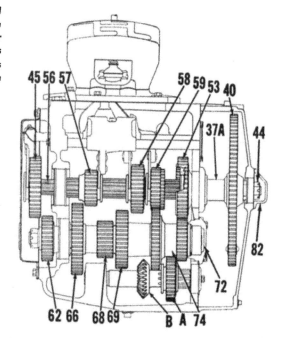

spring, fuel line from carburetor, coil wire from coil and fuel tank support from governor housing. Raise rear of hood approximately 2 inches and block up between hood and governor housing. Remove the cap screws re-

Fig. JD1144—The first reduction gear cover can be loosened from main case by tapping end of drive shaft with a soft hammer.

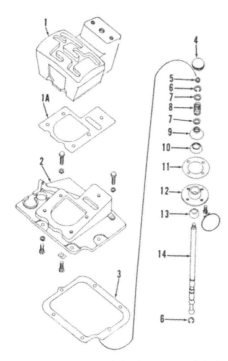

Fig. JD1145—Exploded view of model 60 transmission top cover and shifter quadrant assembly. Spring (8) should require 50 pounds to compress it to a height of 15/16-1 1/16 inches.

1. Quadrant	8. Spring
2. Transmission cover	9. Ball socket cover
3. Gasket	10. Ball socket seal
4. Gear shift ball	11. Gasket
5. Lock washer	12. Fulcrum ball socket
6. Snap ring	13. Fulcrum ball
7. Washer	14. Gear shift lever

taining the shifter quadrant and transmission cover assembly to main case and withdraw assembly from tractor.

132. SHIFTER SHAFTS AND SHIFTERS. To remove the shifter shafts and shifters, first remove the engine flywheel as outlined in paragraph 59 and the transmission top cover as outlined in the preceding paragraph 131. Remove the fourth and sixth speed shifter pawl and spring. They are located in a vertical drilled hole in main case to the left of the top opening and are retained by threaded retainer (28—Fig. JD1146).

Using a pry bar through transmission top opening, move each shifter along its shaft until the detent pawls rise and hold the pawls in the raised position with a cotter pin inserted in the exposed hole of each pawl.

Move the left rear wheel out on axle and remove the fourth and sixth speed gear cover which is located under flywheel on left side of main case.

Remove cap screw (15 — Figs. JD1146 and 1147) and adjusting screw (16) from left end of front and rear shifter shaft. Pull each shifter shaft toward left to disengage it from locking plate (20—Fig. JD1146), rotate shafts sufficiently to move detents out of alignment with pawls, withdraw shifter shafts from left and shifters from above as follows:

Withdraw shaft (17) and shifter (18). Unwire and remove set screw (26). Withdraw shaft (27), shifters (19 and 24), shaft (31) and shifter (25).

New shifter yokes can be riveted to shifters if old yokes are worn or bent. Renew any shifter shaft that is worn around the detents. Renew any pawl that is worn out-of round at ball end.

The first and third as well as the second and fifth speed shifter springs should test 47-53 pounds when compressed to 1⅛ inches. The fourth and sixth speed shifter spring should test 51-63 pounds when compressed to 1⅞ inches.

When reinstalling the shifter shafts and shifters, refer to Fig. JD1146 and reverse the removal procedure, making certain that flat on right end of shifter shafts engage locking plate (20). After the shifter shafts and shifters are installed, place the fourth and sixth speed shifter in neutral position, making certain that pawl (30) engages detent in shaft. Turn adjusting screw (16), located at left end of front and rear shifter shaft, until right hand gates are aligned with gate of shifter on center shaft.

Fig. JD1147—Left side of model 60 main case with fourth and sixth speed gear cover removed.

15. Locking cap screw	45. Fourth and sixth
16. Adjusting screw	speed sliding gear
32. Fourth and sixth	60. Nut
speed shifter	62. Fourth and sixth
33. Set screw	speed drive gear

Fig. JD1146—Exploded view of model 60 shifter shaft and shifters.

15. Locking cap screw
16. Adjusting screw
17. Second and fifth shifter shaft
18. Second and fifth shifter
19. Overdrive shifter
20. Shifter shaft locking plate
22. Shifter pawl
23. Spring
24. Fourth and sixth shifter
25. First, third and reverse shifter
26. Set screw
27. Fourth and sixth shifter shaft
28. Pawl retainer
29. Spring
30. Fourth and sixth shifter pawl
31. First, third and reverse shifter shaft
32. Fourth and sixth speed shifter
33. Set screw

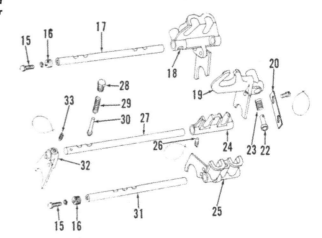

133. SLIDING GEAR SHAFT. To remove the sliding gear shaft (56—Fig. JD1148) and gears, remove the transmission case top cover as in paragraph 131 and the shifter shafts and shifters as outlined in paragraph 132. Remove the sheet metal oil retainer (46—Fig. JD1149) and extract outer snap ring (48) from main case. Withdraw sliding gear shaft from left side of main case and remove gears from above. Pilot bearing (52) can be removed from the transmission drive shaft after removing snap ring (51).

Use Fig. JD1149 as a guide during installation and install the sliding gear shaft by reversing the removal procedure. Install outer snap ring (48) with gap in snap ring spanning the oil passage in main case. Install the sheet metal oil retainer (46) with flat spot adjacent to oil passage in main case.

134. TRANSMISSION DRIVE SHAFT. To remove the drive shaft (53—Figs. JD1148 and 1149), remove the clutch, belt pulley and first reduction gear cover as outlined in paragraph 130. Remove the sliding gear shaft as in the preceding paragraph 133.

Withdraw the first reduction gear (40) and spacer (37A) if tractor is equipped with a transmission driven powershaft, or idler gear (37) if engine driven powershaft is used. Remove bearing cover (35) and withdraw drive shaft from tractor.

When installing the shaft, refer to Fig. JD1149 as a reference and reverse the removal procedure.

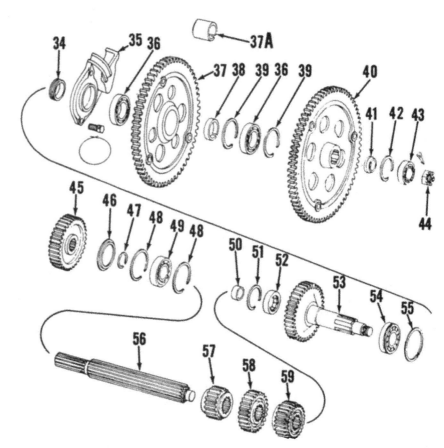

Fig. JD1149—Exploded view of model 60 sliding gear shaft, transmission drive gear and associated parts.

34. Oil collar	43. Bearing	52. Pilot bearing
35. Bearing cover	44. Nut	53. Transmission drive shaft
36. Bearing	45. Fourth and sixth speed	54. Bearing
37. Powershaft idler	sliding gear	55. Snap ring
37A. Spacer	46. Oil retainer	56. Sliding gear shaft
38. Spacer	47. Snap ring	57. First and third sliding
39. Snap ring	48. Snap ring	pinion
40. First reduction gear	49. Bearing	58. Second and fifth sliding
41. Spacer	50. Pilot bearing inner race	pinion
42. Snap ring	51. Snap ring	59. Sliding gear shaft drive gear

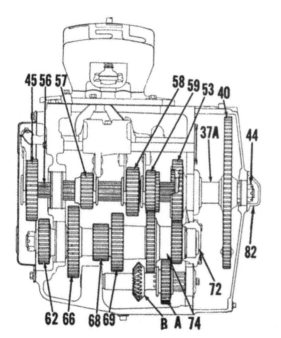

Fig. JD1148 — Rear sectional view of model 60 transmission with non-continuous power take-off. An idler gear replaces spacer (37A) on models equipped with engine driven power shaft.

A. Powershaft sliding gear
B. Powershaft drive pinion
37A. Spacer
40. First reduction gear
44. Nut
45. Fourth and sixth sliding gear
53. Transmission drive shaft
56. Sliding gear shaft
57. First and third sliding pinion
58. Second and fifth sliding pinion
59. Sliding gear shaft drive gear
62. Fourth and sixth drive gear
66. First and third gear
68. Differential drive pinion
69. Second and fifth gear
72. Countershaft
74. Countershaft idler gear

135. COUNTERSHAFT. To remove the transmission countershaft (72—Figs. JD1148 and 1150), remove the sliding gear shaft as in paragraph 133 and the transmission drive shaft as in the preceding paragraph. Remove cotter pin and nut (60) from left end of countershaft and using a suitable puller, remove the fourth and sixth speed drive gear (62). Remove the countershaft right bearing housing (81), bump countershaft out right side of main case and withdraw gears from above.

The right bearing cup can be pulled from bearing housing (81) if renewal is required and the left bearing cup can be driven from the main case bore.

Use Fig. JD1150 as a guide during reassembly and install the same number of shims (80) as were originally removed. Install the fourth and sixth speed drive gear (62) and tighten nut (60) securely. Mount a dial indicator as shown in Fig. JD1151 and check the countershaft end play which should be 0.001-0.004. If the end play is not as specified, remove the countershaft right bearing housing (81 — Fig. JD1150) and add or remove the required amount of shims (80).

Model 70

136. FIRST REDUCTION GEAR COVER. To remove the first reduction gear cover, disconnect the clutch operating rod and drain oil from cover. Remove belt pulley as outlined in paragraph 119. Move the right wheel out, unbolt and remove the right brake assembly. Remove the transmission drive shaft right bearing cover,

extract cotter pin and remove nut from end of transmission drive shaft. Remove cap screws retaining the reduction gear cover to main case and bump end of drive shaft with a soft hammer in a manner similar to that shown in Fig. JD1144 to loosen the reduction gear cover. Pry cover from its locating dowels and withdraw the first reduction gear cover from tractor. Be careful not to lose the camshaft end play removing spring which is retained by the reduction gear cover.

When reassembling, soak new reduction gear cover gasket until gasket is pliable, shellac gasket to main case and install reduction gear cover by reversing the removal procedure. Make certain that oiler gear (47—Fig. JD1152) meshes properly with the first reduction gear before tightening the cover cap screws.

137. TRANSMISSION TOP COVER. To remove the transmission top cover and shifter quadrant assembly, remove grille and loosen the two cap screws retaining front of hood to the radiator top tank. Remove the four cap screws retaining instrument panel to support, disconnect oil pressure line from gage, pull instrument panel rearward and remove the two cap screws retaining the steering shaft and hood support to the gear shift quadrant. Disconnect battery cable and disconnect wires which go through the gear shifter quadrant from instrument panel. On late models so equipped, disconnect oil lines from the automatic fuel shut off valve located at top of fuel filter. On all models, disconnect oil line from connector located at right rear of gov-

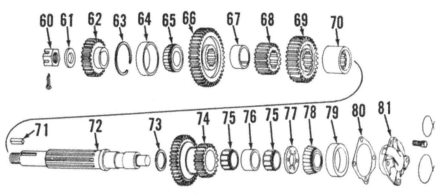

Fig. JD1150—Exploded view of model 60 transmission countershaft. Recommended end play of 0.001-0.004 is adjusted with shims and gaskets (80).

60. Nut	67. Spacer	74. Countershaft idler gear
61. Washer	68. Differential drive pinion	75. Bearing
62. Fourth and sixth speed drive gear	69. Second and fifth speed gear	76. Spacer
	70. Spacer	77. Thrust washer
63. Snap ring	71. Key	78. Bearing cone
64. Bearing cup	72. Countershaft	79. Bearing cup
65. Bearing cone	73. Collar	80. Shims and gaskets
66. First and third speed gear		81. Bearing housing

Fig. JD1151—Using a dial indicator to check end play of model 60 transmission countershaft. Recommended end play of 0.001-0.004 is controlled by shims under the right bearing housing.

60. Nut
62. Fourth and sixth speed drive gear
72. Countershaft

Fig. JD1152—Inside view of model 70 reduction gear cover, showing the installation of the oil slinger gear.

46. Oil slinger gear pin   49. Transmission drive
47. Oil slinger gear   shaft right bearing

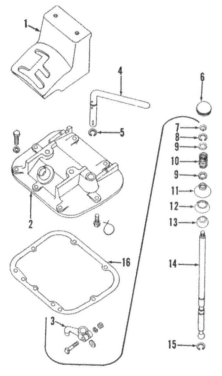

Fig. JD1153—Model 70 transmission cover and shifter quadrant assembly.

1. Quadrant	9. Washer
2. Transmission cover	10. Spring
3. Underdrive shifter lever	11. Fulcrum ball socket cover
4. Underdrive shifter handle	12. Ball socket seal
5. Snap ring	13. Fulcrum ball
7. Lock washer	14. Shift lever
8. Snap ring	15. Snap ring
	16. Gasket

ernor case. Disconnect speed control rod from governor spring, fuel line from carburetor, coil wire from coil and fuel tank support from governor housing and upper water pipe. Raise rear of hood approximately 2 inches and block up between hood and governor housing. Remove tool box. Remove the cap screws retaining the shifter quadrant and transmission cover assembly to main case and withdraw assembly from tractor.

138. SHIFTER SHAFTS AND SHIFTERS. To remove the transmission shifter shafts and shifters, first

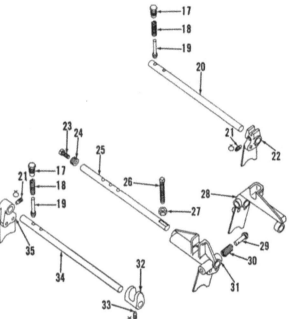

remove the engine flywheel as outlined in paragraph 59 and the transmission top cover as in the preceding paragraph. Move the left wheel out and remove the fifth and sixth speed gear cover which is located on left side of main case. Remove both of the threaded plugs (17—Fig. JD1154) and extract springs (18) and pawls (19) from the vertical drilled holes in main case. Loosen jam nut (27) and set screw (26). Using a pry bar through the transmission top opening, move each shifter along its shaft until the detent pawls rise and hold the

Fig. JD1154 — Exploded view of model 70 transmission shifter shafts and shifters.

17. Pawl retainer
18. Spring
19. Pawl
20. Underdrive shifter shaft
21. Set screw
22. Underdrive shifter
23. Locking cap screw
24. Adjusting screw
25. Shifter shaft
26. Set screw
27. Jam nut
28. Second and reverse shifter
29. Pawl
30. Pawl spring
31. Third and fourth shifter
32. Fifth and sixth shifter arm
33. Set screw
34. Fifth and sixth speed shifter shaft
35. Fifth and sixth speed shifter

pawls in the raised position with a cotter pin inserted in the exposed hole in each pawl. Unwire and remove set screws (21 and 33). Pull each shifter shaft toward left, rotate shafts sufficiently to move detents out of alignment with pawls, withdraw shifter shafts from left and remove shifters from above.

New shifter yokes can be riveted to shifters if old yokes are worn or bent. Renew any shifter shaft that is worn around the detents. Renew any pawl that is worn out-of-round at ball end. Springs (18) should test 51-63 pounds when compressed to $1\frac{7}{8}$ inches. Spring (30) should test 47-53 pounds when compressed to $1\frac{1}{8}$ inches.

When reinstalling the shifter shafts and shifters, refer to Fig. JD1154 as a guide and reverse the removal procedure. Place the underdrive shifter (22) in neutral position, making certain that pawl (19) engages detent in shaft (20). Turn adjusting screw (24), located at left end of shaft (25), until shifter gates are aligned. Tighten set screw (26) and lock with jam nut (27). Install and tighten the locking cap screw (23—Fig. JD1155).

139. SLIDING GEAR SHAFT. To remove the sliding gear shaft (57—Fig. JD1156) and gears, remove the transmission case top cover as outlined in paragraph 137 and the shifter shafts and shifters as in the preceding paragraph 138. Remove the sheet metal oil retainer (61) and extract outer snap ring (58) from main case. Withdraw the sliding gear shaft from left side of main case and remove gears from above. Pilot bearing (51) can be removed from the transmission drive shaft (50) after removing snap ring (52).

Using Fig. JD1156 as a guide, reinstall the sliding gear shaft by reversing the removal procedure. Install outer snap ring (58) with gap in snap ring spanning the oil passage in main case. Install the sheet metal oil retainer (61) with flat spot adjacent to oil passage in main case.

140. TRANSMISSION DRIVE SHAFT. To remove the transmission drive shaft (50—Fig. JD1156), remove the clutch, belt pulley and the first reduction gear cover as outlined in paragraph 136. Remove the sliding gear shaft as in the preceding paragraph 139. Withdraw the first reduction gear (41) and spacer (38A) if tractor is equipped with a transmis-

Fig. JD1155—Left side of model 70 main case with fifth and sixth speed gear cover removed.

Q. Differential bearing quill
20. Underdrive shifter shaft
23. Locking cap screw
34. Fifth and sixth speed shifter shaft
35. Fifth and sixth speed shifter
62. Fifth and sixth speed sliding pinion
65. Fifth and sixth speed drive gear

sion driven powershaft, or idler gear (38) if engine driven powershaft is used. Remove bearing cover (36) and withdraw drive shaft from tractor.

When reinstalling the shaft, refer to Fig. JD1156 as a reference and reverse the removal procedure.

140A. COUNTERSHAFT. To remove the transmission countershaft (79—

Fig. JD1157), remove the sliding gear shaft as in paragraph 139 and the transmission drive shaft as in paragraph 140. Remove cotter pin and nut (63) from left end of countershaft and using a suitable puller, remove the fifth and sixth speed drive gear (65). On models equipped with engine driven (live) power take-off, unbolt pto shifter bearing (B—Fig. JD1158),

pull bearing outward and remove sliding gear (SG). On all models, remove the countershaft right bearing housing (71) and bump countershaft out right side of main case.

NOTE: All of the countershaft gears and spacers except cluster gear (69—Fig. JD1157) can be removed at this time. If the cluster gear is damaged and requires renewal, it will be necessary to perform the following preliminary work in order to provide removal clearance. Drain hydraulic system, remove platform and remove hydraulic lines. Loosen bolt circle retaining final drive housing to main case and separate the housings approximately ½-inch. Remove the differential left bearing quill (Q—Fig. JD1155). This will allow differential to drop down and back enough to provide removal clearance for the cluster gear. Be careful not to damage the gasket when separating the final drive housing from main case.

When reinstalling the countershaft, use the same number of shims (72—Fig. JD1157) as were originally removed and tighten nut (63) securely. Mount a dial indicator with contact button resting on left end of countershaft in a manner similar to that shown in Fig. JD1151 and check the countershaft end play which should be 0.001-0.004. If end play is not as specified, remove the countershaft right bearing housing and add or remove the necessary amount of shims (72).

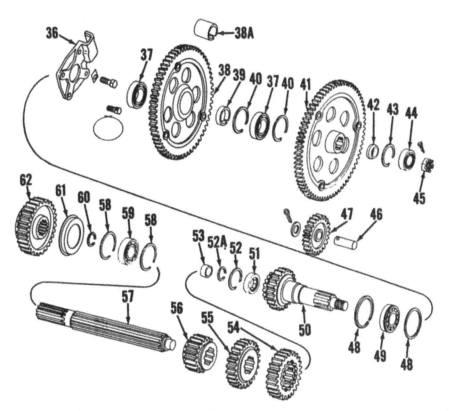

Fig. JD1156—Exploded view of model 70 transmission sliding gear shaft and associated parts. Oil slinger gear (47) is mounted in the first reduction gear cover as shown in Fig. JD1152.

36. Bearing cover	46. Oil slinger gear pin	55. Third and fourth sliding
37. Bearing	47. Oil slinger gear	pinion
38. Powershaft idler	48. Snap ring	56. First and second sliding
38A. Spacer	49. Bearing	pinion
39. Spacer	50. Transmission drive shaft	57. Sliding gear shaft
40. Snap ring	51. Pilot bearing	58. Snap ring
41. First reduction gear	52. Snap ring	59. Bearing
42. Spacer	52A. Snap ring (late models)	60. Snap ring
43. Snap ring	53. Bearing inner race	61. Oil retainer
44. Bearing	54. Sliding gear shaft drive	62. Fifth and sixth sliding
45. Nut	gear	pinion

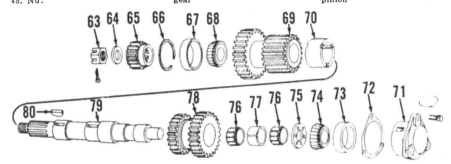

Fig. JD1157—Exploded view of model 70 transmission countershaft. Shaft end play of 0.001-0.004 is controlled by shims and gaskets (72).

63. Nut	68. Bearing cone	75. Thrust washer
64. Washer	69. Cluster gear	76. Roller bearings
65. Fifth and sixth speed	70. Spacer	77. Spacer
drive gear	71. Bearing housing	78. Countershaft idler gear
66. Snap ring	72. Shims & Gaskets	79. Countershaft
67. Bearing cup	73. Bearing cup	80. Woodruff key
	74. Bearing cone	

Fig. JD1158—Right side of model 70 main case with first reduction gear cover removed. The unit is equipped with live power take-off.

B. Pto shifter shaft	36. Drive shaft bearing
bearing	cover
SG. Live pto slid-	71. Countershaft right
ing gear	bearing housing

DIFFERENTIAL, FINAL DRIVE AND REAR AXLE

Fig. JD1159—Model 60 transmission, differential and final drive. Except for detailed differences, models 50 and 70 are similar.

Model 50 tractors are equipped with a two pinion type differential; whereas, the differential in models 60 and 70 is of the three pinion type. The differential unit is mounted on the spider of a spur (ring) gear which meshes with a driving pinion on the transmission countershaft. Pressed through the spider is the differential cross shaft which forms the journals for the integral differential side gears and spur (bull) pinions. The outer ends of the differential cross shaft carry taper roller bearings which support the differential unit. The remainder of the final drive includes the final drive (bull) gears, located in the rear axle housing, and the wheel axle shafts to which the bull gears are splined.

DIFFERENTIAL

Model 50

141. REMOVE AND REINSTALL. To remove the differential, first drain hydraulic system and main case. Remove clutch, belt pulley and both brake assemblies. Remove the first reduction gear cover as in paragraph 123 and withdraw the first reduction

gear and the powershaft idler gear. Remove platform and the hydraulic "Powr-Trol" lines. Remove seat and disconnect battery cable and wiring harness at rear of tractor. Support tractor under main case and unbolt rear axle housing from main case. With the rear axle housing assembly supported so that it will not tip, roll the unit rearward and away from tractor. Remove both of the differential bearing quills (1 and 14—Fig. JD1160) and withdraw the differential assembly, left hand end first.

When reinstalling, use the same thickness of shims and gaskets (2) as were originally removed, mount a dial indicator so that contact button is resting on side of the ring gear and check the differential end play which should be 0.001-0.004. If end play is not as specified, remove the left bearing quill and add or remove the required amount of shims and gaskets (2).

142. OVERHAUL. With the differential removed as outlined in the preceding paragraph, proceed to disassemble the unit as follows: Remove snap rings (4) and using a suitable

puller, remove both of the differential bearing cones (5) and withdraw the combination differential side gears and spur (bull) pinions (7). Remove rivets (10), extract pinion shafts (9) and remove bevel pinions (11).

Neither the differential spider or spur ring gear (8) is available separately. If either part is damaged, renew the entire assembly. Check the disassembled parts against the values which follow:

Thickness of thrust wash-
ers (6) 0.216-0.222
I.D. of pinions (7) 2.069-2.071
Diameter of differential
shaft (12) 2.065-2.066

NOTE: If shaft (12) is worn, press the old shaft out, support differential spider on a piece of pipe and press the new shaft in place as shown in Fig. JD1161.

I.D. of bevel pinions
(11—Fig. JD1160) 0.863-0.866
Diameter of pinion
shafts (9) 0.8585-0.860

When reassembling, reverse the dis-assembly procedure and press bearing cones (5) on the differential shaft until they seat against thrust washers (6). Install new snap rings (4).

Models 60-70

143. **REMOVE AND REINSTALL.** To remove the differential, first drain the hydraulic system and main case.

Remove the platform and the hydraulic "Powr-Trol" lines. Remove seat and disconnect battery cable and wiring harness at rear of tractor. Support tractor under main case and unbolt rear axle housing from main case. With the rear axle housing assembly sup-

ported so that it will not tip, roll the unit rearward and away from tractor. Remove the left brake assembly (or slide it out about one inch), remove the differential left bearing quill (1—Fig. JD1162) and withdraw the differential assembly, right end first, as shown in Fig. JD1163.

When reinstalling, use the same thickness of shims and gaskets (2—Fig. JD1162) as were originally removed, mount a dial indicator so that contact button is resting on side of the ring gear and check the differential end play which should be 0.001-0.004.

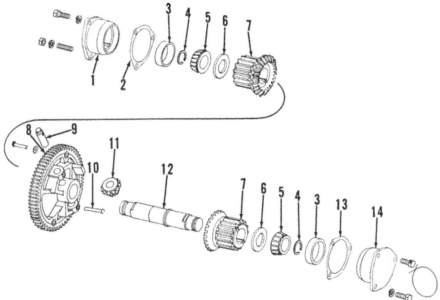

Fig. JD1160—Exploded view of model 50 differential. The spur ring gear and spider are available as an assembled unit only.

1. Left bearing quill
2. Shims and gaskets
3. Bearing cup
4. Snap ring
5. Bearing cone
6. Thrust washer
7. Differential side gear and bull pinion
8. Ring gear and spider
9. Pinion shaft
10. Rivet
11. Bevel pinion
12. Differential cross shaft
13. Gasket
14. Right bearing quill

Fig. JD1163—When removing models 60 and 70 differential, withdraw the right end first. If the differential only is to be removed, it is not necessary to remove both brake assemblies. Refer to text.

Fig. JD1161—When pressing the differential cross shaft in model 50 differential, be sure to support the spider with a piece of pipe as shown. A similar procedure is used on models 60 and 70.

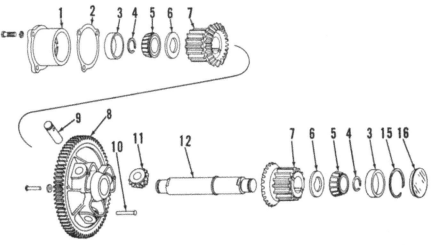

Fig. JD1162—Exploded view of model 60 differential assembly. Model 70 is similar except snap rings (4) are not used. Differential end play is adjusted with shims and gaskets (2).

1. Left bearing quill
3. Bearing cup
5. Bearing cone
6. Thrust washer
7. Differential side gear and bull pinion
8. Ring gear and spider
9. Pinion shaft
10. Rivet
11. Bevel pinion
12. Differential cross shaft
15. Snap ring
16. Bearing cover

If end play is not as specified, remove the left bearing quill and add or remove the required amount of shims and gaskets (2).

144. OVERHAUL. With the differential removed as outlined in the preceding paragraph, proceed to disassemble the unit as follows: On model 60, remove snap rings (4—Fig. JD1162). On all models use a suitable puller and remove the bearing cones (5). Withdraw the combination differential side gears and spur (bull) pinions (7). Remove rivets (10), extract pinion shafts (9) and remove bevel pinions (11).

Neither the differential spider or spur ring gear (8) is available separately. If either part is damaged, renew the entire assembly. Check the disassembled parts against the values which follow:

Thickness of thrust washers (6)
Model 60 ....0.225-0.230
Model 70 ....0.287-0.292
I.D. of pinions (7)
Model 60 ....2.238-2.240
Model 70 ....2.597-2.599
Diameter of differential shaft (12)
Model 60 ....2.234-2.235
Model 70 ....2.592-2.593

NOTE: If shaft (12) is worn, press the old shaft out, support differential spider on a piece of pipe and press the new shaft in place in a manner similar to that shown in Fig. JD1161.

I.D. of bevel pinions (11—Fig. JD1162)
Model 60 ....1.114-1.116
Model 70 ....1.114-1.116
Diameter of pinion shafts (9)
Model 60 ....1.1085-1.1100
Model 70 ....1.1085-1.1100

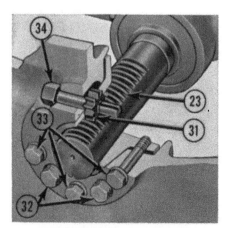

Fig. JD1164—Models 50, 60 and 70 rear wheels can be removed by loosening cap screws (33), turning jack screws (32) in and turning pinion screw (34).

When reassembling, reverse the removal procedure and press bearing cones (5) on the differential shaft until they seat against thrust washers (6). On model 60, install new snap rings (4).

FINAL DRIVE
Models 50-60-70

145. R & R REAR WHEEL. Support rear of tractor, turn wheel until rack in axle shaft is in the up position and unscrew the three cap screws (33—Fig. JD1164) approximately 5/16 inch. Turn the two jack screws (32) clockwise until outer groove in each screw is flush with outer surface of wheel hub. Turn pinion shaft screw (34) and remove wheel.

When reinstalling, make certain both rear wheels are set the same distance from the centerline of the tractor. A slotted hole in back of battery box indicates tractor center line.

146. AXLE SHAFT OUTER FELT SEAL RENEW. To renew the rear wheel axle shaft outer felt seal (21—Fig. JD1165), remove wheel as in the preceding paragraph and using a cold chisel and hammer as shown in Fig. JD1166, drive the felt seal retainer

(22) from the axle housing. Withdraw the seal retainer and felt seal.

When reassembling, renew the seal retainer, and using a brass drift and

Fig. JD1166—Removing outer retainer for model 60 axle shaft outer felt seal. The same procedure is used on model 50 and model 70.

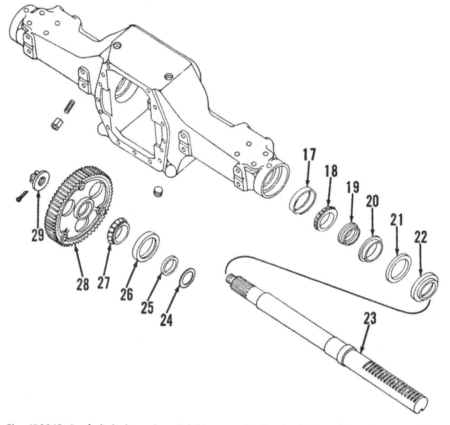

Fig. JD1165—Exploded view of model 50 rear wheel axle shaft and housing assembly. Models 60 and 70 are similar except washer (24) is not used.

17. Bearing cup	21. Felt seal	26. Bearing cup
18. Bearing cone	22. Outer retainer for felt seal	27. Bearing cone
19. Bearing spacer	23. Wheel axle shaft	28. Final drive (bull) gear
20. Inner retainer for felt seal	24. Washer (50 only)	29. Adjusting nut
	25. Inner oil seal	

hammer, drive the retainer in until it seats against recess in housing.

147. FINAL DRIVE (BULL) GEAR, WHEEL AXLE SHAFT & BEARINGS AND/OR AXLE SHAFT INNER OIL SEAL. Drain hydraulic system, transmission and final drive housing. Remove platform, hydraulic lines and seat. Disconnect battery cable and wiring harness at rear of tractor. Support the complete basic housing assembly and attach units in a chain hoist so arranged that the complete assembly will not tip. Unbolt basic housing from rear axle housing and move the complete assembly away from tractor.

CAUTION: This complete assembly is heavy and due to the weight concentration at the top, extra care should be exercised when swinging the assembly in a hoist.

Remove cotter pin and loosen, but do not remove the adjusting nut (29—Fig. JD1165). Using a hammer and a long taper wedge as shown in Fig. JD1167, force axle shaft loose from the bull gear. Remove nut (29) and withdraw gear.

Withdraw axle shaft and inspect housing between inner and outer seals for presence of transmission oil. If oil is found, the inner seal (25—Fig. JD1165) should be renewed. The need for further disassembly is evident.

When reassembling, pack the axle shaft outer bearing with wheel bearing grease. Alternately tighten nut (29) and bump outer end of axle shaft

to assure proper seating of the taper roller bearings. The proper adjustment is when the axle shaft has an end play of 0.001-0.004; then tighten to the nearest castellation and install the cotter pin. When installing the basic housing, turn the power (PTO) shaft to engage the coupling splines.

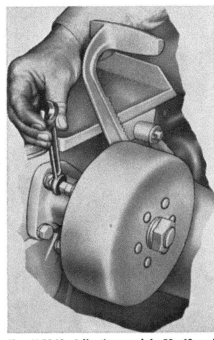

Fig. JD1168—Adjusting models 50, 60 and 70 brakes. Brake pedals should have a free travel of 1¾ inches on model 50, 3 inches on models 60 and 70.

Fig. JD1167—On models 50, 60 and 70, the axle shaft can be forced out of the final drive (bull) gears by loosening the adjusting nut and driving a long tapered wedge between the nuts.

BRAKES

Models 50-60-70

150. ADJUSTMENT. To adjust the brakes, tighten the adjusting screw as shown in Fig. JD1168 to reduce the pedal free travel to approximately 1¾ inches for model 50, 3 inches for models 60 and 70.

151. R & R BRAKE SHOES. To remove the brake shoes for lining replacement, loosen adjusting screw (41—Fig. JD1169), remove nut (56) and using a rawhide hammer, bump brake shaft (37) inward to free drum from shaft. Withdraw the brake drum, pry shoes away from adjusting pins to release spring tension and remove shoes.

Install brake shoes by reversing the removal procedure and adjust the brakes as outlined in paragraph 150.

152. R & R AND OVERHAUL BRAKE ASSEMBLY. Either brake assembly can be removed from model 50 tractors by first removing drum, then unbolting housing assembly from tractor. On model 60 tractors, use a double nut arrangement and remove the upper inside implement mounting stud from the front face of rear axle housing before attempting to unbolt and remove the brake assembly. The removal procedure is evident on model 70 tractors.

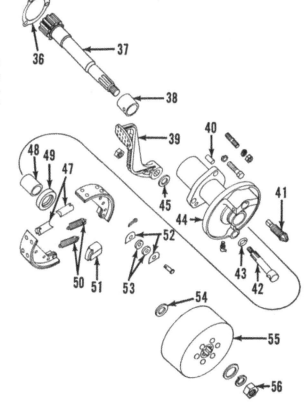

Fig. JD1169 — Exploded view of models 50, 60 and 70 brake. Adjustment is accomplished by turning screw (41).

36. Gasket
37. Brake shaft
38. Inner bushing
39. Pedal
40. Dowel pin
41. Adjusting screw
42. Pedal shaft
43. "O" ring
44. Housing
45. Washer
47. Adjusting pin
48. Outer bushing
49. Dust guard
50. Spring
51. Cam
52. Washer
53. Rollers
54. Thrust washer
55. Brake drum
56. Nut

The procedure for disassembling the brakes is evident after an examination of the unit and reference to Fig. JD1169. Be sure to mark the relative position of pedal shaft (42) with respect to pedal (39) and housing (44). Check the disassembled parts against the values which follow:

I.D. of inner bushing (38)
Model 50 .... 1.627-1.628
Models 60-70 .... 1.7535-1.7540
Diameter of brake shaft at inner bushing
Model 50 .... 1.619-1.620
Models 60-70 .... 1.749-1.750

Fig. JD1170—The outer brake shaft bushing on models 50, 60 and 70 should be installed to a depth of 1/16 inch below end of housing as shown.

I.D. of outer bushing (48)
Model 50 .... 1.3762-1.3767
Models 60-70 .... 1.4982-1.4987
Diameter of brake shaft at outer bushing
Model 50 .... 1.372-1.373
Models 60-70 .... 1.494-1.495

Defective bushings can be removed from the brake housing by sawing through the bushing wall with a hack saw and driving bushing out. When cutting the inner bushing, be careful not to saw the partition in housing just beyond end of bushing.

Install new bushings by using a suitable piloted drift. Outer bushing should be installed to a depth of 1/16 inch as shown in Fig. JD1170.

When reassembling, vary the number of thrust washers (54—Fig. JD1169) to provide a brake shaft end play of 0.004-0.044.

TRANSMISSION DRIVEN POWER TAKE-OFF SYSTEM

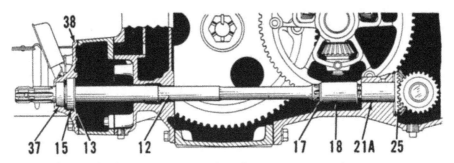

Fig. JD1180—Sectional view showing models 50 and 60 non-continuous (transmission driven) powershaft installation. The general construction of model 70 is similar except bevel gear and shaft (25) is carried in bearings instead of bushings.

12. Powershaft	15. Bearing	18. Coupling	37. Oil seal
13. Snap ring	17. Snap ring	21A. Bushings	38. Bearing housing

The transmission driven (non-continuous) power take-off system receives its drive from an idler gear which is mounted on the transmission countershaft. Power from the transmission countershaft is transmitted through a sliding gear to a shaft and bevel pinion which meshes with the bevel drive gear (25—Fig. JD1180). The power output shaft (12) is connected to bevel gear (25) by coupling (18).

For removal and overhaul of the transmission countershaft, refer to paragraph 128 for model 50, 135 for model 60 and 140A for model 70.

OVERHAUL
Models 50-60-70

160. **POWER (OUTPUT) SHAFT, BEARING & SEAL.** To remove the power (output) shaft bearing (15—Fig. JD1180) and/or seal (37), first remove the powershaft shield and unbolt bearing housing (38). Pull powershaft and bearing housing assembly rearward far enough to extract the internal snap ring (13). The bearing housing can be removed at this time and the bearing and/or seal renewed. To remove the powershaft, it is necessary to perform the additional work of removing the basic housing (rear axle housing cover) as follows:

Drain hydraulic system, transmission and final drive housing. Remove platform, hydraulic lines and seat. Disconnect battery cable and wiring harness at rear of tractor. Support the complete basic housing assembly and attached units in a chain hoist so arranged that the complete assembly will not tip. Unbolt basic housing from rear axle housing and move the complete assembly away from tractor.

CAUTION: This complete assembly is heavy and due to the weight concentration at the top, extra care should be exercised when swinging the assembly in a hoist.

When reassembling, reverse the disassembly procedure and use tin sleeve or shim stock over the powershaft splines to avoid damage when installing oil seal (37).

161. BEVEL GEARS AND SHIFTER. To overhaul the power take-off driving bevel gears and associated parts, first drain the hydraulic system and main case. Remove the platform and the hydraulic "Powr-Trol" lines. Remove seat and disconnect battery cable and wiring harness at rear of tractor. Support tractor under main case and unbolt rear axle housing from main case. With the rear axle housing assembly supported so that it will not tip, roll the unit rearward and away from tractor.

Remove the transmission countershaft as in paragraph 128 for model 50, 135 for model 60 or 140A for model 70.

Working through top of main case, remove cotter pin (2—Figs. JD1181 or 1182) from shifter lever and pull the lever part way out of the transmission case. Using a drift, bump shifter arm (33) down and off the shaft. Remove Woodruff key (34) snap ring (35), spring (1) and withdraw shifter lever from above. Refer to Fig. JD1183.

On models 50 and 60, remove cotter pin (3B—Fig. JD1181) and screw the shifter shaft part way out of main case and withdraw shifter (32D) from shifter shaft, catching ball (6) and spring (5) as they fly out.

On model 70, remove cap screws (4—Fig. JD1182) retaining bearing (3) to main case, withdraw bearing (3) and shifter (32).

On all models, remove bearing cover (8), bump drive shaft (29E or 30—Figs. JD1181 or 1182) toward right until bearing (10) emerges from case; and using a suitable puller, remove bearing from shaft. Withdraw drive shaft and pinion through top opening in main case. Reach down through top opening in main case and withdraw bevel gear and shaft (25) from the main case boss as shown in Fig. JD1185.

On models 50 and 60, the drive shaft and bevel gear shaft are carried in bushings in the main case bosses as shown in Fig. JD1185. Refer to Fig. JD1181 and check the shafts and bushings against the values which follow:

I.D. of bevel gear bushings (21A)

Models 50 & 60 ....1.377-1.379

I.D. of drive shaft bushing (27F)

Model 50 ....1.502-1.504
Model 60 ....1.377-1.379

O.D. of bevel gear shaft at bushing (21A)

Models 50 & 60 ....1.370-1.371

Fig. JD1183—Top view of transmission case with powershaft shifter lever partially removed. Arm (33) must be bumped off lever.

  1. Spring   35. Snap ring
34. Woodruff key   36. Shifter lever

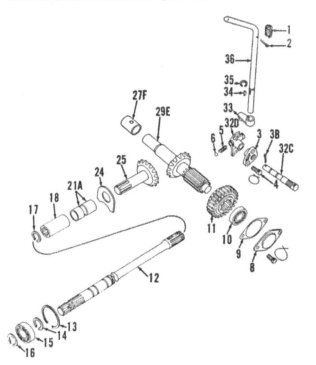

Fig. JD1181—Exploded view of model 50 non-continuous (transmission driven) powershaft. Model 60 is similar. Bevel gear mesh position is controlled by the thickness of thrust washer (24) and backlash is controlled by shims (9).

Fig. JD1182—Exploded view of model 70 non-continuous (transmission driven) powershaft. Bevel gear and shaft bearings are retained by cover (19). Drive shaft bearing is retained by snap rings (26 & 28).

1. Spring	8. Bearing cover
2. Cotter pin	9. Shims
3. Shifter shaft bearing	10. Bearing
3B. Cotter pin	11. Sliding gear
4. Cap screws	12. Powershaft
5. Detent spring	13. Snap ring
6. Detent ball	14 & 16. Snap ring
7. Safety wire	15. Bearing

17. Snap ring (some models)	24. Thrust washer
18. Coupling	25. Bevel gear & shaft
19. Bearing cover (70)	26 & 28. Snap rings (70)
20. Gaskets (70)	27. Bearing (70)
21 & 23. Bearings (70)	27F. Bushing (50 & 60)
21A. Bushings (50 & 60)	29. Bevel pinion (70)
22. Spacer (70)	29E. Pinion and shaft (50 & 60)

30. Drive shaft (70)
31. Woodruff key (70)
32. Shifter & shaft (70)
32D. Shifter (50 & 60)
33. Shifter arm
34. Woodruff key
35. Snap ring
36. Shifter lever

O.D. of drive shaft at bushing (27F)

Model 50 ....1.499-1.500
Model 60 ....1.370-1.371

When installing the drive shaft bushing, make certain that oil hole in bushing is in register with oil hole in the bushing boss. When installing the bevel gear shaft bushings, make certain that bushings do not cover oil hole in the bushings boss. The front bushing should extend forward of the bushing boss by 0.082 or approximately three fourths the thickness of the fibre thrust washer. The rear bushing should be flush with rear of bushing boss. Install bushings with narrow end of oil grooves toward oil hole in boss. Groove in front bushing should be in the 9 o'clock position when viewed from rear and groove in rear bushing

should be in the 12 o'clock position when viewed from rear.

NOTE: Some very early tractors were equipped with one bushing for the bevel gear shaft. In all probability these tractors were changed over to the two bushing construction; however, if the one bushing construction is encountered, it is important that two bushings be installed. Before installing the two new bushings, drill a ⅜ inch hole through the bushing boss 1 13/32 inches forward of rear face of bushing boss. This will place the oil hole between the new bushings when they are installed properly.

On model 70, the drive shaft and bevel gear shaft are carried in bearings as shown in Fig. JD1182. The drive shaft bearing (27) is retained in the case boss by snap rings (26 and

28). The bevel gear shaft bearings (21 and 23) are retained in the case boss by bearing cover (19).

On all models, when installing the bevel gear and shaft, make certain that smooth side of thrust washer (24) is toward the bevel gear and that the washer fits into the machined recess on front of the case boss. On models 50 and 60, this thrust washer controls the mesh position of the bevel gears and should not be excessively worn. Thickness of models 50 and 60 washer is 0.102-0.106. On model 70, vary the number of shims (20—Fig. JD1182) so that heels of bevel gears are in register. When installing the drive shaft, vary the number of shims (9—Figs. JD1181 or JD1182) between bearing cover (8) and main case to give the bevel gears a backlash of 0.006-0.010.

Fig. JD1184—Right view of models 50 and 60 main case with the powershaft shifter shaft partially removed.

Fig. JD1185 — Cut-away view of models 50 and 60 main case showing removal of the power-shaft bevel gear and shaft. The fibre thrust washer fits into the machined recess (R) at front of bushing boss.

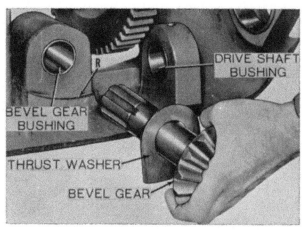

ENGINE DRIVEN POWER TAKE-OFF SYSTEM

The continuous (engine driven) power take-off system receives its drive from the right end of the engine crankshaft. To visualize the power train, refer to Fig. JD1187.

The power shaft drive gear (DG) is keyed to the engine crankshaft and is in constant mesh with a spur idler gear which is carried on two ball bearings mounted on the transmission drive gear and shaft. When sliding gear (SG) is shifted into mesh with the idler gear (IG), power is transmitted rearward to a multiple disc, wet type clutch (C) via a pair of bevel gears in the bottom of the transmission case. Engagement of the clutch permits power to flow through a pair of spur gears to the power take-off (output) shaft (S).

CLUTCH, OUTPUT SHAFT & OIL PUMP

Models 50-60-70

165. **ADJUST CLUTCH.** To adjust the clutch on model 50 tractors prior 5017822, 60 tractors prior 6033676 and 70 tractors prior 7010776, proceed as follows: Remove the clutch inspection hole plug, engage the clutch and check the distance the clutch cam disc has moved back from the brake plate as shown in Fig. JD1188. The desired clearance is 0.090 and can be checked with a No. 13 gage steel wire as shown. If the adjustment is incorrect, remove the oil filler plug and note the position of the bent-over tang on the cam locking washer. If tang is not in register with oil filler opening, shift the pow-

er shaft drive out of mesh and turn the power (output) shaft until the bent over tang is accessible. Disengage clutch and bend tang out of notch in adjusting cam.

To tighten clutch, turn the adjusting cam in a counter-clockwise direction one notch, engage the clutch and check the clearance between the cam disc and brake plate. Continue this procedure until the adjustment is correct when checked in two or three positions. When adjustment is complete, bend the lock washer tang into one of the notches in the adjusting cam. Remove the clutch fork adjusting screw cap nut (Fig. JD1188), engage the clutch and turn the adjusting screw in until it just touches the clutch fork:

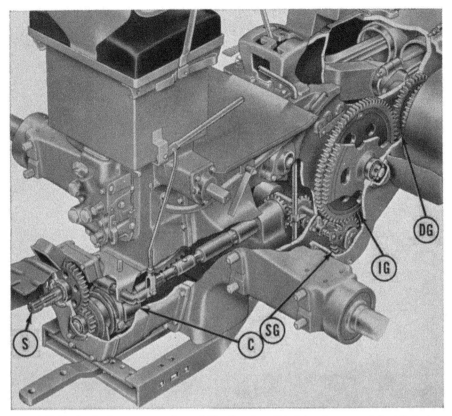

Fig. JD1187—Typical cut-away view of the live (engine driven) power take-off system on models 50, 60 and 70. Drive gear (DG) is keyed to the engine crankshaft.

C. Clutch   IG. Idler gear   S. Power (output) shaft   SG. Sliding gear

then, back the screw up one complete turn. Tighten the jam nut and reinstall the cap nut.

166. The adjusting procedure on 50 tractors after 5017821, 60 tractors after 6033675 and 70 tractors after 7010775 is similar to the earlier models except the cam locking washer is not used. To unlock the cam for adjustment purposes, turn the set screw

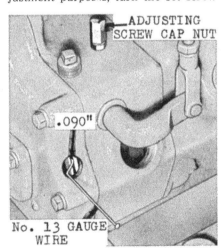

Fig. JD1188—Desired adjustment of models 50, 60 and 70 powershaft clutch is when the clutch cam disc has moved 0.090 back from the brake plate when clutch is engaged. The clearance can be checked with a No. 13 gage steel wire as shown.

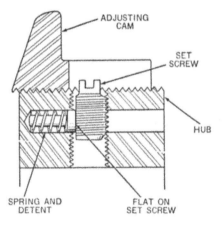

Fig. JD1190 — Removing 50, 60 and 70 live pto clutch, housing cover and output shaft assembly. Pushing down on the fork shaft (24) will facilitate separating the cover from clutch housing.

(Fig. JD1189) into hub until head of screw clears the cam. To lock the cam after adjustment is complete, back the set screw out until its head engages one of the slots in the adjusting cam.

167. **OVERHAUL CLUTCH AND OUTPUT SHAFT.** To remove the clutch and/or power (output) shaft, first drain oil from clutch housing and engage the clutch. Back off the clutch fork adjusting screw (S—Fig. JD1190), remove the powershaft guard and disconnect the clutch operating linkage. Remove the cap screws retaining the clutch housing cover to the clutch housing, push down on the clutch fork shaft (24) and withdraw clutch, cover and powershaft assembly as shown.

NOTE: The clutch drum cannot be serviced at this time. If the drum requires renewal, refer to paragraph 168.

167A. Inspect the clutch fork and shaft (Fig. JD1191). The parts can be removed after removing cotter pin (P).

Support the removed clutch and housing cover assembly in a soft jawed vise and using two large screw drivers, disengage the clutch as shown in Fig. JD1192. Remove snap ring (38) retaining clutch plates to clutch shaft and withdraw clutch plates, discs, release springs, adjusting cam, balls and clutch collar. Unlock the brake plate retaining cap screws, unscrew the screws evenly to avoid damaging the brake plate and withdraw cam with disc, springs and washers. Note the exact number and location of the springs so they can be reinstalled in the same position. If there are shims under the brake plate, save them for reinstallation.

Fig. JD1189—Live powershaft clutch adjusting cam and hub as used on late model 50, 60 and 70 tractors. The unit is locked with a set screw.

Fig. JD1191—Models 50, 60 and 70 clutch housing with cover and clutch plates removed.

P. Cotter pin
22. Clutch fork
24. Fork shaft
40. Clutch drum

Fig. JD1192—Disengaging the live pto clutch, using two large screw drivers. Removal of snap ring (38) will permit disassembly of the clutch.

Remove bearing cover (29—Fig. JD1193), extract snap ring (27) from rear end of clutch shaft and bump or press clutch shaft (12) out of bearing (26). Unbolt and remove oil seal housing (1), remove snap ring (5) and bump or press the powershaft (7) out of bearing (9). NOTE: On some late model tractors, the powershaft (7) is carried in taper roller bearings. On earlier models, inspect the bushing in clutch housing which carries the front end of the powershaft. If clearance between bushing and shaft is excessive, renew the worn part.

Shaft diameter at bush-
ing 1.367-1.368
I.D. of bushing 1.370-1.372
Desired running clear-
ance 0.002-0.005

Outer pressure springs (14 — Fig. JD1193) should require 83-93 pounds to compress them to a height of 1⅜ inches. Test specifications for the inner springs are not available.

On models 50 and 60, the clutch return springs (34) should require 12¼-15 pounds to compress them to a height of 1 3/16 inches. On model 70, the springs should require 12.5-15.5 pounds to compress them to a height of 1 21/32 inches.

Inspect clutch shaft and pilot bushing (39) which is located in clutch drum.

I.D. of bushing (39) ....0.8102 Min.
Shaft dia. at bushing
(39) 0.8082-0.8092
Desired running clear-
ance 0.001-0.002

If running clearance is excessive, renew the worn part. NOTE: On late model tractors and for replacement purposes on earlier models, the pilot bushing and thrust washer are two pieces.

Inspect the clutch facings to make certain they are in good condition and be sure plates and discs are not worn or out-of-flat more than 0.008.

When reassembling, use thin shim stock around the powershaft splines to avoid damage when installing oil seal and housing (1). It is advisable to coat the seal with gun grease prior to installation.

To reassemble the clutch, proceed as follows: With the assembled clutch housing cover supported in a soft jawed vise, install washer (13), inner and outer springs (14) in their original position and cam (16) with disc (15). If there were shims (25) installed between the brake plate and housing, install the same quantity as were removed. Install brake plate and turn the cap screws down evenly, a little at a time, to avoid bending or breaking the brake plate. When the cap screws are tight, lock them in position with the tab washers. Install clutch collar (19), balls (17), adjusting cam with

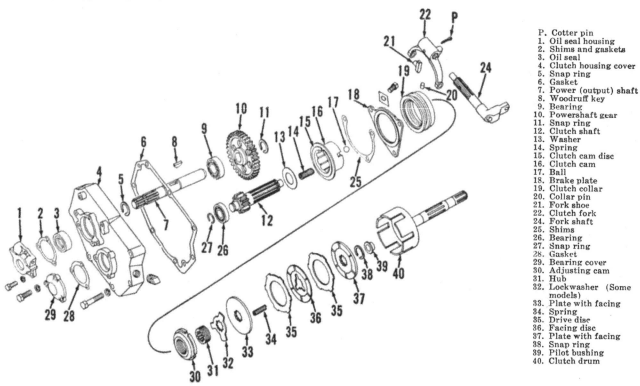

P. Cotter pin
1. Oil seal housing
2. Shims and gaskets
3. Oil seal
4. Clutch housing cover
5. Snap ring
6. Gasket
7. Power (output) shaft
8. Woodruff key
9. Bearing
10. Powershaft gear
11. Snap ring
12. Clutch shaft
13. Washer
14. Spring
15. Clutch cam disc
16. Clutch cam
17. Ball
18. Brake plate
19. Clutch collar
20. Collar pin
21. Fork shoe
22. Clutch fork
24. Fork shaft
25. Shims
26. Bearing
27. Snap ring
28. Gasket
29. Bearing cover
30. Adjusting cam
31. Hub
32. Lockwasher (Some models)
33. Plate with facing
34. Spring
35. Drive disc
36. Facing disc
37. Plate with facing
38. Snap ring
39. Pilot bushing
40. Clutch drum

Fig. JD1193—Exploded view of early production live power take-off clutch. The power (output) shaft (7) on some late models is carried in taper roller bearings.

hub (31) and on early models so equipped, install locking washer (32). NOTE: The adjusting cam on late models is locked by a set screw (Fig. JD1189). Refer also to Fig. JD1194.

Install plate (33—Fig. JD 1193) with facing up, followed by a steel drive disc (35). Install a facing disc (36), a steel drive disc (35), alternating in this order until all lined and unlined plates are installed. Install springs (34). Install the thick driven plate (37) with facing down and install snap ring (38).

The clutch can be adjusted by using a feeler gage as shown in Fig. JD1195 and referring to the general procedure given in paragraph 165; or, the clutch can be adjusted after installation on tractor by following the procedure in paragraph 165.

Before installing clutch, use a spare clutch drum or straight edge, align clutch plates and engage clutch. Install clutch and housing cover assembly by reversing the removal procedure.

168. OVERHAUL CLUTCH DRUM AND OIL PUMP. To remove the clutch drum, first drain hydraulic system, clutch housing and main case. Remove clutch and clutch housing cover assembly as outlined in paragraph 167. Remove platform, hydraulic lines and seat. Disconnect battery cable and wiring harness at rear of tractor. Support the complete basic housing assembly and attached units in a chain hoist so arranged that the complete assembly will not tip. Unbolt basic housing from rear axle housing and move the complete assembly away from tractor.

CAUTION: This complete assembly is heavy and due to the weight concentration at the top, extra care should be exercised when swinging the assembly in a hoist.

Remove the cap screws retaining the clutch oil pump body to front side of basic housing and remove pump body and idler gear as shown in Fig. JD1196. Remove cotter pin (P—Fig. JD1197) and withdraw clutch fork shaft (24) and fork (22). Pull clutch drum (40) from basic housing.

On early models so equipped, examine clutch drum and bushing (42—Fig. JD1198) for being excessively worn.

Bushing inside diameter 1.3745-1.3755
Drum shaft dia. at bushing (42) 1.366-1.368
Desired running clearance 0.0065-0.0095

On other models, the drum shaft is supported in an anti-friction bearing.

Fig. JD1194—Assembling the engine driven power take-off clutch on early models equipped with the cam locking washer (32).

17. Ball
18. Brake plate
19. Clutch collar
30. Adjusting cam

Fig. JD1196—Engine driven power shaft clutch oil pump body and idler gear removed from front face of basic housing. The pump drive gear and shaft is also the clutch drum shaft.

Fig. JD1195—It is possible and often more convenient to adjust the pto clutch prior to final assembly on the tractor. In which case, a feeler gauge can be used to check the adjustment.

Fig. JD1197—Models 50, 60 and 70 clutch housing with cover and clutch plates removed.

P. Cotter pin
22. Clutch fork
24. Fork shaft
40. Clutch drum

The front end of the clutch drum shaft is supported by bushing (46). The bushing and/or drum shaft should be renewed if clearance is excessive.

I.D. of bushing 1.370-1.372
Drum shaft dia. at bushing (46) 1.366-1.368
Desired running clearance 0.002-0.006

Check the clutch oil pump parts against the values which follow:

O.D. of idler gear 2.115-2.117
I.D. of idler gear bore in hsg. 2.121-2.123
O.D. of idler gear shaft 0.9994-1.0000
I.D. of idler gear bushing 1.002-1.003
Thickness of idler gear 0.997-0.998
Depth of idler gear bore in hsg. 1.000-1.005

When reassembling, coat mating surfaces of pump body and basic housing with shellac and use shim stock around splines of drum shaft to avoid damaging the pump housing oil seal. Refer to Fig. JD1199.

DRIVING GEARS, BEVEL GEARS & SHIFTER

Models 50-60-70

169. **RENEW DRIVE AND IDLER GEARS.** To renew the powershaft drive gear (DG—Fig. JD1200) and/or idler gear (IG), first remove engine clutch, belt pulley and first reduction gear cover as outlined in paragraph 123 for model 50, paragraph 130 for model 60 or paragraph 136 for model 70. Withdraw spacer (S), first reduction gear and powershaft idler gear. Remove snap ring (SR) and using a suitable puller as shown in Fig. JD1201, remove the drive gear.

The idler gear is carried on two anti-friction type bearings which should be inspected and renewed if their condition is questionable.

When reassembling, install drive gear, using a brass drift as shown in Fig. JD1202.

170. **OVERHAUL BEVEL GEARS & SHIFTER.** To overhaul the power take-off driving bevel gears and associated parts, first drain the hydraulic system and main case. Remove the platform and the hydraulic "Powr-Trol" lines. Remove seat and disconnect battery cable and wiring harness at rear of tractor. Support tractor under main case and unbolt rear axle housing from main case. With the rear axle housing assembly supported so that it will not tip, roll the unit rearward and away from tractor.

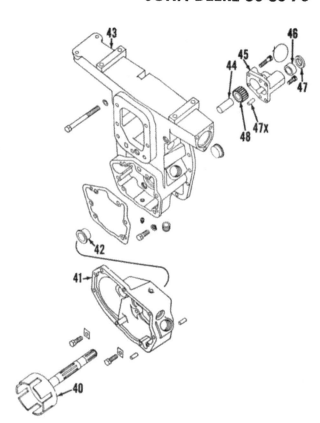

Fig. JD1198—Typical basic housing, clutch oil pump and associated parts used on models 50, 60 and 70. On some models, the drum shaft is carried in a bearing instead of bushing (42).

40. Clutch drum
41. Clutch housing
42. Bushing (some models)
43. Basic housing
44. Idler gear shaft
45. Pump housing
46. Bushing
47. Oil seal
47X. Dowel pin
48. Idler gear

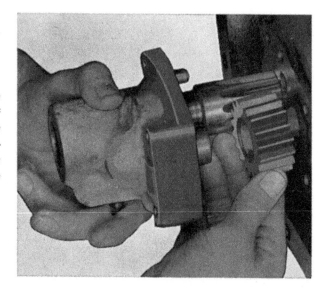

Fig. JD1199—Assembling the live power take-off clutch oil pump. Mating surfaces of pump housing and basic housing should be coated with shellac.

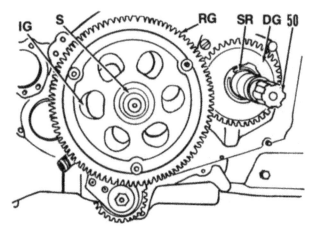

Fig. JD1200—Right side of model 60 main case with reduction gear cover removed. Models 50 and 70 are similar.

DG. Power shaft drive gear
IG. Power shaft idler gear
RG. First reduction gear
S. Spacer
SR. Snap ring
50. Crankshaft

Fig. JD1201—Using puller to remove the continuous power shaft drive gear on model 50 tractor. The same procedure can be used on models 60 and 70.

Fig. JD1202 — Using a hammer and brass drift to install the continuous power shaft drive gear on model 50. The same procedure can be used on models 60 and 70.

Remove the transmission countershaft as in paragraph 128 for model 50, 135 for model 60 or 140A for model 70.

Working through top of main case, remove cotter pin (2—Fig. JD1204 or 1205) from shifter lever and pull the lever part way out of the transmission case. Using a drift, bump shifter arm (33) down and off the shaft. Remove Woodruff key (34), snap ring (35), spring (1) and withdraw shifter lever from above. Refer to Fig. JD1206. Remove snap ring or cotter pin (27—Fig. JD1204 or 1205), slide spool (3)

Fig. JD1206—Top view of transmission case with power shaft shifter lever partially removed. Arm (33) must be bumped off lever.

1. Spring	35. Snap ring
34. Woodruff key	36. Shifter lever

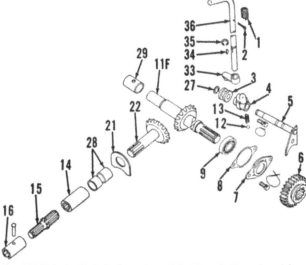

Fig. JD1204—Exploded view of models 50 and 60 engine driven powershaft driving bevel gears and associated parts. A cotter pin replaces snap ring (27) on late models.

Fig. JD1205—Exploded view of model 70 engine driven powershaft driving bevel gears and associated parts.

1. Spring	8. Shims	13. Detent spring
2. Cotter pin	9. Bearing	14. Coupling
3. Spool	10. Woodruff key (70)	15. Coupling shaft
4. Shaft bearing	11. Drive shaft (70)	16. Coupling
5. Shifter shaft	11F. Drive gear & shaft	17. Bearing cover (70)
6. Sliding gear	(50 & 60)	18. Gaskets (70)
7. Bearing cover	12. Detent ball	19. Bearing (70)

20. Spacer (70)	27. Cotter pin or snap ring
21. Thrust washer	28. Bushings (50 and 60)
22. Bevel gear and shaft	29. Bushing (50 and 60)
23. Snap ring	33. Shifter arm
24. Bearing	34. Woodruff key
25. Bevel pinion (70)	35. Snap ring
26. Snap ring (70)	36. Shifter lever

from shifter shaft and unbolt shifter shaft bearing from main case. Remove shifter shaft (5), bearing (4) and sliding gear (6).

Remove bearing cover (7), bump drive shaft (11 or 11F) toward right until bearing (9) emerges from case and using a suitable puller, remove bearing from shaft. Withdraw drive shaft and pinion through top opening in main case. Reach down through top opening in case and withdraw bevel gear and shaft (22) from main case boss as shown in Fig. JD1207.

On models 50 and 60, the drive shaft and bevel gear shaft are carried in bushings in the main case bosses as shown in Fig. JD1207. Check the shafts and bushings against the values which follow:

I.D. of bevel gear bushings (28)
    Models 50 & 60 . . . . . . .1.377-1.379
I.D. of drive shaft bushing (29)
    Model 50 . . . . .1.502-1.504
    Model 60 . . . . .1.377-1.379
O.D. of bevel gear shaft at bushing (28)
    Models 50 & 60 . . . . . .1.370-1.371
O.D. of drive shaft at bushing (29)
    Model 50 . . . . .1.499-1.500
    Model 60 . . . . .1.370-1.371

When installing the drive shaft bushing, make certain that oil hole in bushing is in register with oil hole in bushing boss. When installing the bevel gear shaft bushings, make certain that bushings do not cover oil hole in the bushing boss. The front bushing should extend forward of the bushing boss by 0.082 or approximately three-fourths the thickness of the fibre thrust washer. The rear bushing should be flush with rear of bushing boss. Install bushings with narrow end of oil grooves toward oil hole in boss. Groove in front bushing should be in the 9 o'clock position when viewed from rear and groove in rear bushing should be in the 12 o'clock position when viewed from rear.

NOTE: Some very early tractors were equipped with one bushing for the bevel gear shaft. In all probability these tractors were changed over to the two bushing construction; however, if the one bushing construction is encountered, it is important that two bushings be installed. Before installing the two new bushings, drill a ⅜ inch hole through the bushing boss 1 13/32 inches forward of rear face of bushing boss. This will place the oil hole between the new bushings when they are installed properly.

On model 70, the drive shaft and bevel gear shafts are carried in bearings as shown in Fig. JD1205. The drive shaft bearing (24) is retained in the case boss by snap rings (23). The bevel gear shaft bearings (19) are retained in the case boss by bearing cover (17).

On all models, when installing the bevel gear and shaft, make certain that smooth side of thrust washer (21) is toward the bevel gear and that the washer fits into the machined recess on front of case boss. On models 50 and 60, the thrust washer controls the mesh position of the bevel gears and should not be excessively worn. Thickness of models 50 and 60 washer is 0.102-0.106. On model 70, vary the number of shims (18—Fig. JD1205) so that heels of bevel gears are in register. When installing the drive shaft, vary the number of shims (8) between cover (7) and main case to give the bevel gears a backlash of 0.006-0.010.

On some early models having the snap ring (27—Fig. JD1204), it is recommended that the snap ring be discarded and the shifter shaft drilled to accommodate a 3/16 x 1 inch cotter pin.

"POWR-TROL" (Hydraulic Lift)

As shown in Fig. JD1210, the hydraulic power lift system is composed of four basic units: The pump unit (P) which is mounted on the governor; the valve housing (VH) which is mounted on the rear face of the basic housing, the rockshaft (RS) and operating cylinder; and the remote control cylinder (RC).

Fig. JD1207 — Cut-away view of models 50 and 60 main case showing removal of the power-shaft bevel gear and shaft. The fibre thrust washer fits into the machined recess (R) at front of bushing boss.

NOTE: The maintenance of absolute cleanliness of all parts is of utmost importance in the operation and servicing of the hydraulic system. Of equal importance is the avoidance of nicks or burrs on any of the working parts.

LUBRICATION AND BLEEDING
Models 50-60-70

180. It is recommended that the "Powr-Trol" working fluid (same weight oil as used in the engine crankcase) be changed at least twice-a-year. Drain housing and flush same with distillate. Fill the valve housing reservoir, start engine, operate the "Powr-Trol" lever several times to bleed the system and fill the remote control cylinder (if used); then refill the reservoir.

Capacity of system is approximately 6 U.S. qts. if remote cylinder is not used; or, 7 U.S. qts. if remote cylinder is used.

Oil recommendations are as follows:
Temp. above 90 Deg. F. . .S.A.E. 30
Temp. 32 Deg. F to
    90 Deg. F. . . . . .S.A.E. 20-20W
Temp. below 32 Deg. F. . .S.A.E. 10W

TROUBLE SHOOTING
Models 50-60-70

181. The following trouble shooting paragraphs often save considerable time in pin pointing malfunctions in the hydraulic system.

The procedure for remedying many of the causes of trouble is evident. The following paragraphs will list the most likely causes of trouble, but only the remedies which are not evident.

181A. **LIFT WILL NOT RAISE.** Probable causes:
a. Low oil supply
b. Wrong weight of oil
c. Faulty pump
d. Relief pressure too low
e. Pump not engaged
f. Relief valve stuck open

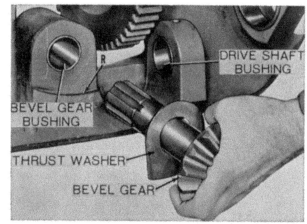

g. Faulty piston ring

h. Load too heavy

i. Gasket failed at pressure passage

j. Inner steel ball missing

k. Bleed hole clogged in upper check valve

l. Sand hole in pressure passage

m. Rockshaft cylinder gaskets leaking

n. Remote cylinder gaskets leaking

181B. LIFT WILL NOT STAY IN RAISED POSITION. Probable causes:

a. "O" rings leaking at lower hose adapter

b. Check valves leaking

c. Piston ring leaking

d. Sand hole in cylinder casting

e. Remote cylinder gaskets leaking

181C. LIFT WILL NOT DROP. Probable causes:

a. Inner ball missing in lower check valve

b. Woodruff key in control shaft sheared

c. Faulty implement

181D. CONTROL LEVER WILL NOT LOCK. Probable causes:

a. Release valve stuck

b. Bleed hole plugged

c. Worn cam

181E. CONTROL LEVER WILL NOT RELEASE. Probable causes:

a. Cam latch roller locked, due to operating valve adjusting screw worn or out of adjustment

b. Worn cam latch locks

c. Release valve stuck. A thorough cleaning will correct this

Fig. JD1211 — Exploded view of models 50, 60 and 70 hydraulic pump. The unit is mounted on the governor housing as shown in Fig. JD1210.

1. Copper washer
2. Plug
3. Housing
4. Dowel pin
5. Idler gear shaft retaining pin
6. Gasket
7. "O" ring
8. Shifter
9. Yoke
10. Washer
11. Drive shaft outer bushing
12. "O" ring
13. Thrust washer
14. Bushings
15. Idler gear with bushings
16. Idler gear shaft
17. Oil seal
18. Drive gear
19. Bushing
20. Drive shaft
21. Woodruff key
22. Pump body
23. Driver gear
24. Bushing
25. Cover with bushing
26. Follower gear
27. Follower gear shaft
28. Dowel pin
29. Set screw
30. Shifter handle
31. Copper washer

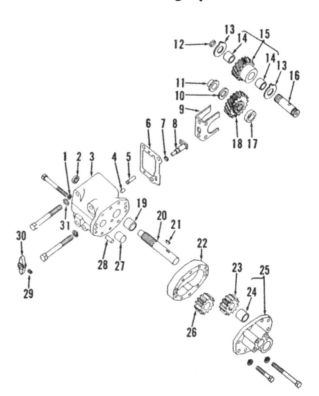

181F. SYSTEM OVER-HEATING. Probable causes:

a. Operator holding lever in raise or lower position after cylinder has reached limit of travel. School operator on proper procedure.

b. Low oil supply

c. Wrong weight oil

d. Relief valve pressure too high

181G. SYSTEM LOSING OIL. Probable causes:

a. Gaskets, "O" rings or seals leaking. Visually check all components of "Powr-Trol."

b. Sand hole or fracture in castings

PUMP UNIT
Models 50-60-70

182. **R & R AND OVERHAUL.** To remove the hydraulic pump, first drain hydraulic system and remove platform. Remove hydraulic lines and unbolt pump from governor case. Withdraw pump being careful not to lose pin (5—Fig. JD1211) which retains idler gear shaft (16) in governor case.

Unbolt and remove pump cover; at which time, the need and procedure for further disassembly will be evident.

After pump is disassembled, wash all parts in solvent and thoroughly examine them for damage or wear. Inspect mating surfaces of housing (3), body (22) and cover (25) for wear or scoring. If follower gear shaft (27) is damaged, it can be removed by welding a piece of metal to it and pulling it out. Inspect all bushings and renew any which show wear.

Pump specifications are as follows:

Gear diameter 2.448-2.450
I.D. of pump body 2.452-2.454
Clearance between gears
& body 0.002-0.006
Gear thickness 0.903-0.904

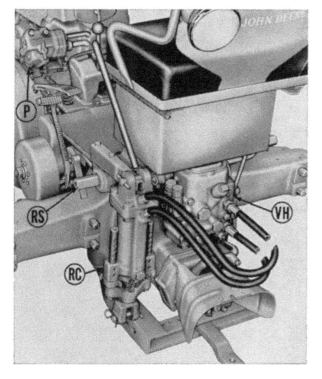

Fig. JD1210—Rear view of John Deere tractor showing the installation of the hydraulic pump (P), "Powr-Trol" valve housing (VH), remote cylinder (RC) and rock-shaft (RS).

Body thickness 0.906-0.907
Clearance between gears
  & cover 0.002-0.004
Follower gear shaft
  diameter 0.9994-1.0000
Follower gear inside di-
  ameter 1.002-1.003
O.D. of drive shaft
  outer end 0.8082-0.8092
I.D. of drive shaft outer
  bushing 0.812-0.813
I.D. of other drive
  shaft bushings 1.002-1.003
O.D. of drive shaft at
  center and inner
  bushings 0.996-0.998
O.D. of idler gear shaft
  (16) 0.9994-1.0000
I.D. of idler gear bush-
  ings (14) 1.002-1.003

When reassembling, renew all "O" rings and seals. Install sliding gear with protrusion toward pumping gears. Coat mating surfaces of housing, body and pump cover with shellac and tighten the cover cap screws securely.

When installing pump, make certain that copper washers (1 & 31) are in good condition.

ADJUST ROCKSHAFT FAST DROP, SLOW RISE AND "POWR-TROL" METERING SCREW
Models 50-60-70

183. **FAST DROP.** To adjust the rockshaft "fast drop," remove cap nut, loosen jam nut and turn the throttle valve screw (73—Fig. JD1212) "in" to increase or "out" to decrease speed of fast drop.

NOTE: Adjustment of the throttle valve depends on the weight of the implement. The valve should not be allowed to seat or rockshaft will not lower.

184. **SLOW RISE.** To adjust the rockshaft "slow rise," remove cap nut, loosen jam nut and turn the metering screw (83) "out" to decrease or "in" to increase speed of slow rise.

185. **METERING SCREW.** To adjust the "Powr-Trol" metering screw so that system can be used for remote cylinder operation, remove cap nut, loosen jam nut and turn the metering screw (83) "in" until it seats firmly.

RELIEF VALVE PRESSURE TEST
Models 50-60-70

186. To check and adjust the relief valve cracking pressure, mount a pressure gage of sufficient capacity (at least 2000 psi) as shown in Fig. JD1213.

NOTE: The shut off valve must be located between the gage and the valve housing on the oil return line. With the shut off valve open, move the "Powr-Trol" operating lever to raise position, allowing the working fluid to circulate through the tube and

Fig. JD1213—Checking operating pressure on John Deere "Powr-Trol" system. The shut-off valve must be located between the pressure gage and the valve housing on the return line.

Fig. JD1212—Models 50, 60 and 70 "Powr-Trol" metering screw assembly (B), throttle valve screw assembly (C) and operating valve adjusting plug and shims assembly (A) exploded from the valve housing.

69. Plug
70. Cap nut
71. Washers
72. Jam nut
73. Throttle valve screw
78. Adjusting washers
83. Metering screw

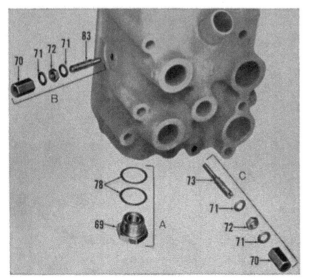

shut off valve and back to the valve housing. Close the valve slowly, and note the pressure reading as the relief valve opens. The relief valve cracking pressure should be 1170-1210 psi.

If the gage reading is not as specified, remove the relief valve spring plug (58—Fig. JD1214) and add washers (60) to increase pressure; or remove washers to decrease pressure. Each washer represents approximately 35 psi.

If the specified gage pressure cannot be obtained, look for a failed or badly worn pump unit. Refer to paragraph 182 for overhaul of the pump.

CHECK VALVES
Models 50-60-70

187. **LEAK CHECK.** Using a spare check valve screw plug, pressure gage, old inner tube valve stem and fittings, make up a leak detector as shown in Fig. JD1215. Remove the upper check valve screw plug (69—Fig. JD1223) from the valve housing, install the leak detector and close off the upper hose adaptor opening. Ap-

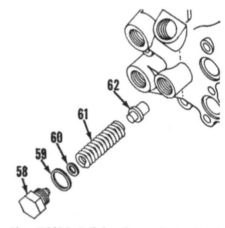

Fig. JD1214—Relief valve and associated parts exploded from models 50, 60 and 70 "Powr-Trol" valve housing. See Fig. JD1223 for legend.

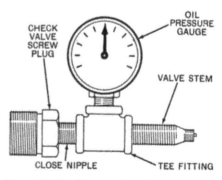

Fig. JD1215—Home-made detector used for checking leaks in models 50, 60 and 70 "Powr-Trol" valve housing check valves.

ply air pressure to the valve stem (30 psi is sufficient) and observe the pressure gage to see if pressure is maintained. Check the lower check valve in the same manner except close off the lower hose adaptor opening.

If the check valves will not hold air pressure, remove the valves and thoroughly clean them. If check valves are leaking, renew the balls and/or seats. Install check valves as outlined in the following paragraph.

188. INSTALL CHECK VALVES. Refer to Fig. JD1216. Install inner check valve with short relief end toward outside which will permit one relief passage to open as shown.

189. ADJUSTMENT. To adjust either of the inner check valves, remove plug (69—Fig. JD1223) and valve lock spring. Insert valve lock (the lock is available from John Deere) in the check valve plug and reinstall the plug in the valve housing. Tighten the plug until both the inner and outer check valves are against their seats

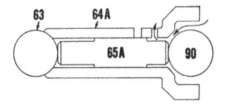

Fig. JD1216—Models 50, 60 and 70 check valve. Correct installation of the inner check valve (65A) permits one relief passage to open in outer check valve.

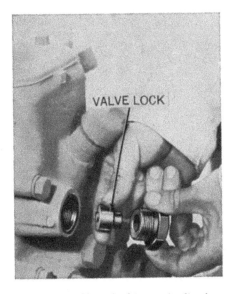

Fig. JD1217—When checking and adjusting John Deere "Powr-Trol" check valves, it is necessary to obtain the valve lock from John Deere.

as shown in Fig. JD1218. At this time, there should be at least ¼ inch movement of the "Powr-Trol" lever when moving from neutral to raise position. If there is less than ¼ inch movement, grind off about 0.001 from inner end of inner check valve and recheck the lever movement.

With the inner check valve adjusted, remove the check valve plug, valve lock and inner check valve. Reinstall the check valve plug and valve lock, leaving out the inner check valve. Tighten the plug until outer check valve is against its seat as shown in Fig. JD1219. Remove the valve housing cover and move the control lever until the cam blade comes against the cam follower roller (21—Fig. JD1223). There should be ⅜-½ inch movement of control lever from this point until the cam on the operating valve tightens against steel ball and outer check valve. If movement is less than ⅜ inch, loosen the cam blade lock screw and move blade **in.** If lever movement is more than ½ inch, move the cam blade **out.** CAUTION: Hold blade firmly against milled slot when making the adjustment as in Fig. JD1220.

Repeat the above procedure, for checking and adjusting the other check valve.

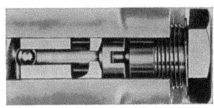

Fig. JD1218—Cut-away view of Deere "Powr-Trol" valve housing showing the valve lock installed for checking the inner check valve.

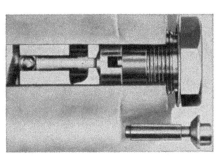

Fig. JD1219—Cut-away view of Deere "Powr-Trol" valve housing showing the valve lock installed for checking the outer check valve.

OPERATING VALVE

Models 50-60-70

190. ADJUST. To adjust the position of the operating valve, place control lever in neutral position and remove the valve housing cover. Place operating valve travel gage (the travel gage is available from John Deere) in the position shown in view (A—Fig. JD1221), with shoulder of gage stem against top of operating valve. Move control lever to lowered position; at which time, the end of the gage stem should just slide over the end of the operating valve as shown in view B. If the opposite end of operating valve strikes plug (69—Fig. JD1212) before end of travel gage stem slides over end of operating valve, add washers (78), which are available in thicknesses of 0.020 and 0.032. If the operating valve travels too far, remove washers to obtain correct travel.

Move control lever to neutral position and set the travel gage as shown in view (B—Fig. JD1221) with end of gage just touching end of operating valve. Move control lever to raise position and gage stem to left; at which time, the shoulder on gage stem should just slide over end of operating valve as shown in view (C). If operating valve has traveled too far, or not enough, adjust stop screw as shown in Fig. JD1222 until desired travel is obtained.

VALVE HOUSING

Models 50-60-70

191. R & R AND OVERHAUL. To remove the "Powr-Trol" valve housing, drain hydraulic system and re-

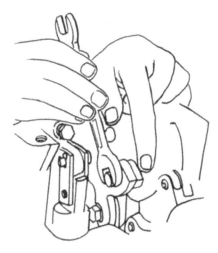

Fig. JD1220 — Adjusting cam follower blades on John Deere "Powr-Trol" valve housing.

move cap screws and nut securing the assembly to basic housing. Withdraw valve housing assembly, being careful not to lose throttle valve (81—Fig. JD1223) and spring (82). On some individual cases, it is necessary to remove the powershaft shield, left cap screw from the housing top cover and/or lower plug (69—Fig. JD1223) before removing the housing.

191A. To disassemble the unit, remove cover, oil line adapters and pipe plugs from housing. Unscrew check valve caps (69) and remove copper washers and springs (67). Remove inner and outer check valves and balls (63, 64 & 65). Remove lock screw from cam forging, drive out control shaft until Woodruff key centers in slot in cam holder; then, drift out the Woodruff key as shown in Fig. JD1224. NOTE: On some models, there is no hole provided in the control shaft for drifting out the Woodruff key. In

which case, bump shaft toward right, viewed from rear until Woodruff key falls out. Withdraw control shaft.

Compress release valve spring and cam follower lever spring with screw drivers and lift out cam and operating valve. Withdraw the follower lever spring. Extract cotter pins holding cam follower lever in place, unscrew

pivot pin screw and remove pivot pin and follower lever. Extract cotters from cam latch pivot pin and remove pin. Remove screws from retainer (28—Fig. JD1223) and withdraw cam latch arm and spring. Remove release valve (31). Remove relief valve plug (58) and extract washers (60), spring (61) and relief valve (62). Remove

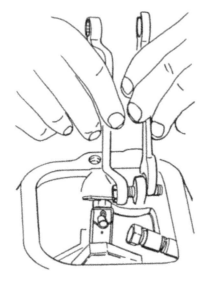

Fig. JD1222—Adjusting John Deere "Powr-Trol" stop screw which controls the operating valve travel from neutral to raise position.

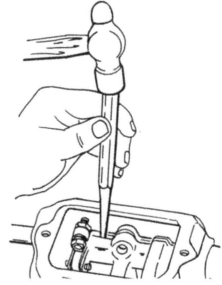

Fig. JD1224—Removing Woodruff key which positions John Deere "Powr-Trol" control shaft in the cam forging.

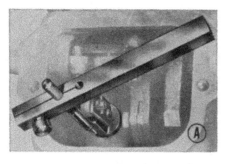

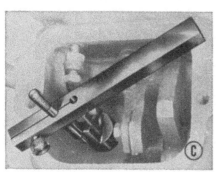

Fig. JD1221—John Deere special travel gage positions for checking the "Powr-Trol" operating valve.

Fig. JD1223 — Exploded view of a typical "Powr-Trol" valve housing assembly as used on models 50, 60 and 70.

1. Holder
2. Centering cam
7. Cam
11. Cam follower arm with roller
13. Pin
14. Link
15. Rivet
16. Operating valve with link
17. Spring
18. Cotter pins
19. Cam blade
20. Pin
21. Roller
23. Cam latch pin
24. Cam latch roller
25. Cam latch arm with roller
28. Retainer
29. Spring
30. Latch rod
31. Release valve
32. Seal
38. Pivot
39. Cap screw
58. Relief valve spring plug
59. Gasket
60. Adjusting washers
61. Spring
62. Relief valve
63. Ball
64. Outer check valve
65. Inner check valve
67. Spring
68. Washer
69. Plug
70. Cap nut
71. Washers
72. Jam nut
73. Throttle valve screw
78. Adjusting washers
81. Throttle valve
82. Spring
83. Metering screw
85. Woodruff key
86. Valve housing
87. Oil seal

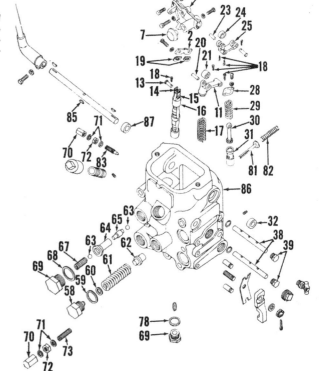

operating valve plug (69) and adjusting washers (78)—save the washers for reinstallation. Remove cap nuts (70), jam nuts (72), by-pass metering screw (83) and throttle valve screw (73).

Wash housing and all removed parts in distillate, inspect all parts and renew any which are questionable. Always renew all "O" rings and seals.

When reassembling, use Fig. JD1223 as a general guide, reverse the disassembly procedure and check and adjust the components as in paragraphs 183, 184, 185, 186, 187, 188, 189 and 190.

ROCKSHAFT & OPERATING CYLINDER
Models 50-60-70

192. **RENEW PISTON SEALS.** To renew the piston seals, first drain hydraulic system and remove the valve housing assembly as in paragraph 191. Extract wire retaining the crank arm pins (2—Fig. JD1225) and withdraw the pins. Remove the rockshaft bearings (14) and pull rockshafts out of basic housing. Withdraw crank arm (4) and rod (5). Apply air pressure at hole (W—Fig. JD1226) and force piston out of cylinder. Examine piston for scoring and renew "O" ring seal (8—Fig. JD 1225) and back up washer (7).

Fig. JD1225 — Exploded view of models 50, 60 and 70 basic housing, rockshaft and rockshaft operating cylinder.

1. Basic housing
2. Pins
3. Cotter pin
4. Crank arm
5. Rod
6. Piston
7. Backing washer
8. "O" ring
9. Gasket
10. Cylinder
11. Rivet
12. Plug
13. Rockshaft
14. Bearing

193. **RENEW CYLINDER.** To renew cylinder (10—Fig. JD1225), first drain hydraulic system and main case. Remove platform, hydraulic lines and seat. Disconnect battery cable and wiring harness at rear of tractor. Support the complete basic housing assembly and attach units in a chain hoist so arranged that the complete assembly will not tip. Unbolt basic housing from rear axle housing and move the complete assembly away from tractor.

CAUTION: This complete assembly is heavy and due to the weight concentration at the top, extra care should be exercised when swinging the assembly in a hoist.

Unbolt cylinder from front face of basic housing. Inspect cylinder for scores, cracks, etc.

When installing the cylinder, tighten the retaining cap screws to a torque of 150 Ft.-Lbs.

REMOTE CYLINDER (FIXED STOP TYPE)
Models 50-60

194. **OVERHAUL.** Remove oil lines and end cap (12—Fig. JD1227). Remove the piston retaining nut and piston assembly, taking care not to damage the Neoprene piston rings. Withdraw piston rod and yoke from cylinder. Remove piston rod stop (19), adjusting rods (18), seal retainer (17) and "V" seal assembly (15).

Inspect face of end cap for nicks or burrs, adjusting rods for being bent and piston rod for burrs, scratches

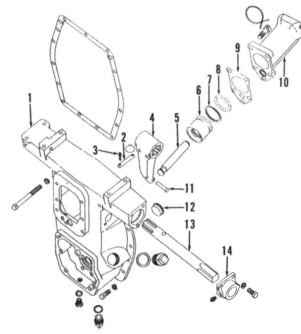

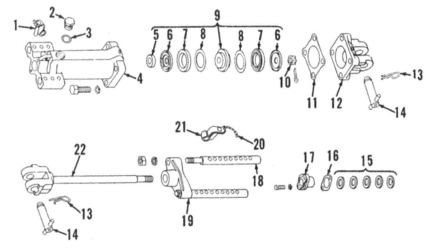

Fig. JD1227—Exploded view of models 50 and 60 fixed stop type remote control cylinder.

1. Stop pin	6. Packing retainer	12. End cap	18. Stop rod
2. Inspection cap	7. "U" cup packing	13. Locking pin	19. Stop
3. Gasket	8. Backing washer	14. Attaching pin	20. Chain
4. Cylinder	9. Piston with seals	15. "V"—seal assembly	21. Stop-pin & clip
5. Spacer	10. Nut	16. Shims	22. Rod
	11. Gasket	17. Oil seal retainer	

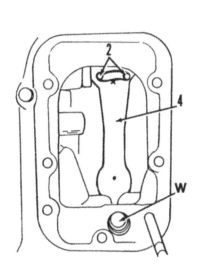

Fig. JD1226—Rear view of models 50, 60 and 70 basic housing with the "Powr-Trol" valve housing removed.

Fig. JD1228—Checking pull required to move lubricated piston rod through "V"—seal assembly on Deere fixed stop type remote control cylinder.

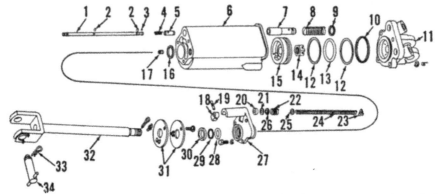

Fig. JD1229—Exploded view of models 50, 60 and 70 hydraulic stop type remote control cylinder.

1. Stop rod	10. Gasket	18. Stop rod arm
2. Snap rings	11. End cap	19. Groov pin
3. Ball	12. Packing washer	20. Packing adapter
4. Bleed valve spring	13. "O" ring	female
5. Bleed valve	14. Nut	21. "V" packing
6. Cylinder	15. Piston	22. Packing spring
7. Stop valve	16. Piston rod guide	23. Stop rod washer
8. Stop valve spring	gasket	24. Stop rod spring
9. Gasket	17. Pipe plug	25. Stop rod washer

26. Packing adapter male
27. Piston rod guide
28. "O" ring
29. Packing washer
30. Wiper seal
31. Rod stops
32. Piston rod
33. Locking pin
34. Attaching pin

and/or being bent. Bent adjusting rods will be O.K., providing they can be thoroughly straightened. Small burrs and/or scratches can be removed from piston rod by using a fine hone; however, piston rod should be renewed if it is bent. Renew any other questionable parts.

Lubricate the piston rod and reassemble the cylinder, leaving out the Neoprene piston rings. Attach a spring scale as shown in Fig. JD1228 and check the pull required to move the lubricated piston rod through the "V" seal assembly. Add or deduct shims (16—Fig. JD1227) until a pull of approximately 4 pounds is required. Shims are 0.010 and 1/32 inch thick. When adjustment is as specified, install the Neoprene piston rings and tighten the piston retaining nut. Using a new gasket, install the end cap and tighten the retaining cap screws to a torque of 85 ft.-lbs.

REMOTE CYLINDER (HYDRAULIC STOP TYPE)
Models 50-60-70

195. **OVERHAUL.** To disassemble the unit, remove oil lines and end cap (11—Fig. JD1229). Remove stop valve (7) and bleed valve (5) by pushing stop rod (1) completely into cylinder. Withdraw stop valve from bleed valve, being careful not to lose the small ball (3). Remove nut from piston rod, being careful not to distort the rod and remove piston and

rod. Push stop rod (1) all the way into cylinder and drift out Groov pin (19). Remove piston rod guide (27).

Examine all parts for being excessively worn and renew all seals. Wiper seal (30) should be installed with sealing lip toward outer end of bore. Install stop rod ("V" seal assembly) (20, 21, 22 & 26) with sealing edge toward cylinder. Complete the assembly by reversing the disassembly procedure and install the piston rod stop as shown in Fig. JD1230.

Fig. JD1230—Models 50, 60 and 70 hydraulic stop remote control cylinder showing the proper installation of the stops.

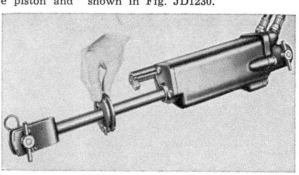